A KEY TO EGYPTIAN GRASSES

A KEY
TO EGYPTIAN GRASSES

Thomas A. Cope

Royal Botanic Gardens, Kew

and

Hasnaa A. Hosni

Cairo University Herbarium

Cairo
University
Herbarium

Royal
Botanic Gardens
Kew

General Editor of Series J.M. Lock

Cover Design by Media Resources, RBG, Kew

ISBN 0 947643 35 4

Typeset at Royal Botanic Gardens, Kew by Pam Arnold,
Christine Beard, Brenda Carey, Margaret Newman,
Pam Rosen and Helen Ward

Printed & Bound in Great Britain by Whitstable Litho, Whitstable, Kent

CONTENTS

DEDICATION

This book is dedicated to TAC's Egyptian 'minders':

Ahmad Gamal El-Deen Fahmy
Ibrahim Ahmed El-Garf
Mohammed El-Gibaly

ACKNOWLEDGEMENTS

Our grateful thanks are due to the authorities of the University of Cairo and the Royal Botanic Gardens, Kew for generously co-sponsoring the project which began with four very interesting and enlightening weeks spent in Egypt by TAC in Spring 1990. We must further acknowledge the authorities of the Agricultural Museum, Cairo (CAIM) and the herbarium of the African University at Aswan —especially Dr. Irina Springuel — for further help and facilities. TAC's special gratitude must be extended to the dedicatees of this book for their unceasing kindness, help and generosity during his stay in Egypt. He could not have managed without them.

INTRODUCTION

by T.A. Cope

It is now half a century since the appearance of Volume 1 (Monocotyledons) of Täckholm & Drar's *Flora of Egypt* (1941), and in the normal course of events a Flora will have lived out its useful life after such a period. Thoughts must then be turned to a revision incorporating not only those many taxonomic and nomenclatural changes that are bound to have taken place in the meantime, but also to the many new finds that are equally likely to have occurred. Efforts by Vivi Täckholm to do just this are embodied in *The Student's Flora of Egypt*, especially in the second edition of 1974, in a much reduced and simplified format. But even this book is now showing its age, particularly when considered against the background of a period of rapidly advancing understanding of grass systematics that has culminated in Clayton & Renvoize's *Genera Graminum* (1986). Furthermore, with the recent appearance of the grass volume for the *Flora of Libya* (Sherif & Siddiqi, 1988) and with work steadily progressing at Kew on a grass volume for the proposed Flora of the Arabian Peninsula, now would seem an opportune time to take a fresh look at the grasses of Egypt.

It was at the suggestion of Prof. Nabil El-Hadidi of the Cairo University Herbarium (CAI) that I embarked on this project. I was very fortunate to have been offered the assistance of Dr Hasnaa A. Hosni, also of CAI, a dedicated student of Egyptian grasses whose contribution included not only the distribution and ecology of species within Egypt, but certain key questions concerning my own taxonomy and key-writing. I was delighted that she agreed to be my co-author. I was further delighted to have been offered the exceptional artistic talents of Magdy El-Gohary whose illustrations have immeasurably enhanced the value of this book.

The format of this key is straightforward and follows in basic design my *Key to the Grasses of the Arabian Peninsula* (1985). It is always to be regretted that grasses are a family that do not lend themselves well to treatment at a 'popular' level, but the keys have been kept as simple as possible, calling for the absolute minimum of dissection and measurement of cryptic and difficult characters — a good × 10 handlens being all that is required for most of the time. On occasion, of course, it is quite impossible to avoid difficult characters if the key is to do its job properly, and in this respect I have had to be quite uncompromising. It is freely admitted that this key is best used by those who are prepared to take the trouble to learn those few terms that are necessary for an understanding of how the grass plant, and in particular its spikelet, is put together. Adequate accounts will be found in Bor (1968) and Clayton & Renvoize (1986), so there is no need of repitition here.

A key to tribes is included, but certain anomalous genera will either defy attempts at placement or persist in coming out in the wrong place. The inflorescence of *Lygeum*, for example, is at first totally baffling; *Pennisetum clandestinum* simply has no visible characters that will lead the key-user to the right place although it has a quite unmistakable habit; *Tetrapogon* will always key to Eragrostideae instead of Cynodonteae, but will be found in the keys to genera of both tribes.

The sequence of tribes and genera follows that of Clayton & Renvoize (1986). The keys to species have been expanded so that the final statement of each dichotomy comprises the following:

1

The key character(s) for determination of species
An additional statement on the habit or peculiarities of the species if this is
    thought appropriate or helpful
Habitat(s) occupied by the species in Egypt
Distribution within Egypt
Brief summary of range beyond Egypt
The name of the plant. The most recent name is given in the text; all
    synonyms, many misapplied names, and authorities will be found in
    Appendix 2.

Distribution within Egypt follows the scheme devised by El-Hadidi (1980):

M: Mediterranean coastal belt
    Mma: the western coastal region from Rashid to the Libyan frontier
    ('Marmarica')
    Md: the Nile Delta region between Rashid and Dumyat
    Mp: the eastern coastal region from Dumyat to Sinai ('Pelusiac-Tanitic')

N: the Nile Valley
    Nv: the valley north of Aswan Province
    Nn: the Nubian Nile of Aswan Province

D: the Deserts
    Dl: the Libyan Desert
    Dn: the Nubian Desert
    Dg: the Galala Desert
    Da: the Arabian Desert
    Di: the Isthmic Desert

O: the Oases of the Lybian (Dl) and Nubian (Dn) Deserts (none in Dg, Da or
    Di)

S: the mountainous region of Sinai proper between the Gulfs of Suez and
    Aqaba

R: the Red Sea coastal plains, north of Sa and including Sinai

Sa: the Sahelian scrub region in and around Gebel Elba

Uw: the massif of Gebel Uweinat

A record in parentheses () signifies a literature record not yet confirmed amongst available herbarium collections; an asterisk (*) denotes an introduction to Egypt or to the botanical district indicated.

No attempt is made in this study to indicate relative abundance of species in Egypt or in its botanical divisions; in many cases this simply is not known. Reference is made in the keys from time to time to an appendix (Appendix 1) where particular taxonomic or nomenclatural problems are discussed in more detail.

All grasses found in Egypt — wild or cultivated, crops or ornamentals — are included as any of them could — and indeed some do — escape and integrate themselves into the native flora. Bamboos are, however, excluded. They require a special study all of their own and the ideal material for naming them is not always available. Their taxonomy is currently going through a period of intensive study and is consequently in a very unsettled state. The conditions under which they grow are so special that it is unlikely that any will ever become naturalised in Egypt. On that slender basis the authors feel justified in excluding them from the present account.

SYNOPSIS OF CLASSIFICATION

Tribe I. ORYZEAE

A widespread tribe of aquatic grasses linked to the bamboos by their leaf-anatomy. The spikelet is 3-flowered with the 2 lower florets reduced to small chaffy scales (1-flowered in 2. *Leersia*) and the glumes to no more than obscure lips at the tip of the pedicel.

    Genera: 1. *Oryza*
                 2. *Leersia*

Tribe II. EHRHARTEAE

A tribe of rather indefinite affinity, but apparently linked to I. Oryzeae by the 2 sterile lemmas below the fertile, and its bambusoid embryo.

    Genus: 3. *Ehrharta*

Tribe III. LYGEAE

A tribe of one species with an extraordinary spikelet. The inflorescence comprises a single spikelet enclosed by a spatheole; there are no glumes and the two lemmas are fused below along their opposing margins to form a cylindrical, coriaceous tube; the paleas are fused back to back below to form a septum in the lemma-tube.

    Genus: 4. *Lygeum*

Tribe IV. STIPEAE

The single indurated lemma without rhachilla-extension distinguishes this tribe which is otherwise closely related to V. Poeae and VIII. Aveneae.

    Genera: 5. *Stipa*
                 6. *Oryzopsis*

Tribe V. POEAE

A large, typically temperate tribe usually recognised by its paniculate inflorescence (except 8. *Lolium*), several-flowered spikelets exserted from the glumes and 5-nerved (3-nerved in 16. *Sphenopus* and 19. *Cutandia*, and apparently so in 17. *Desmazeria*), laterally compressed lemmas.

    Genera: 7. *Festuca*
                 8. *Lolium*
                 9. *Vulpia*
              10. *Cynosurus*
              11. *Lamarckia*
              12. *Briza*

13. *Poa*
14. *Dactylis*
15. *Eremopoa*
16. *Sphenopus*
17. *Desmazeria*
18. *Catapodium*
19. *Cutandia*
20. *Ammochloa*

Tribe VI. HAINARDIEAE

A small tribe closely related to V. Poeae, characterised by its cylindrical, fragile racemes with the spikelets sunk in cavities and covered by the collateral glumes.

Genus: 21. *Parapholis*

Tribe VII. MELICEAE

Although closely related to V. Poeae, members of this tribe are united by their connate sheath-margins and different basic chromosome number. The rhachilla terminates in a clavate clump of several sterile lemmas (in the Egyptian representative).

Genus: 22. *Melica*

Tribe VIII. AVENEAE

A large temperate tribe characterised by long glumes of a rather scarious texture (shorter and herbaceous in V. Poeae) and geniculate dorsal awn, but the latter is variable in its occurrence. The spikelets may be several-flowered (subtribe Aveninae), 3-flowered with the 2 lower florets reduced to barren chaffy scales at the base of the fertile (subtribe Phalaridinae), or strictly 1-flowered (subtribe Alopecurinae)

Subtribe AVENINAE
Genera: 23. *Avena*
24. *Trisetaria*
25. *Rostraria*
26. *Holcus*
27. *Corynephorus*

Subtribe PHALARIDINAE
Genus: 28. *Phalaris*

Subtribe ALOPECURINAE
Genera: 29. *Agrostis*
30. *Ammophila*
31. *Triplachne*
32. *Gastridium*
33. *Lagurus*
34. *Polypogon*
35. *Alopecurus*
36. *Phleum*

Tribe IX. BROMEAE

A tribe resembling V. Poeae, but with a curious hairy appendage on the ovary; it also has the peculiar rounded starch-grains typical of X. Triticeae.

Genera: 37. *Bromus*
         38. *Boissiera*

Tribe X. TRITICEAE

This tribe can usually be recognised by its racemose inflorescence. It is allied to V. Poeae, but the endosperm contains simple starch-grains of a distinctive rounded shape, and the ovary is surmounted by a hairy appendage.

Genera: 39. *Brachypodium*
         40. *Elymus*
         41. *Taeniatherum*
         42. *Crithopsis*
         43. *Hordeum*
         44. *Agropyron*
         45. *Eremopyrum*
         46. *Triticum*
         47. *Aegilops*

Tribe XI. ARUNDINEAE

A heterogeneous tribe whose Egyptian members fall into 3 groups: those with long glumes and bilobed lemmas (genera 48 & 49); those exhibiting dioecy (genus 50); and those with reed-like habit (genera 51 & 52).

Genera: 48. *Schismus*
         49. *Centropodia*
         50. *Cortaderia*
         51. *Arundo*
         52. *Phragmites*

Tribe XII. ARISTIDEAE

The spikelet in this tribe contains a single floret whose lemma is tightly rolled around the grain at maturity. The lemma is surmounted by 3 awns, one (or rarely all three) of them sometimes hairy. In many species the awns are connate at the base into a twisted column. All members of the tribe are characteristic of dry, hostile environments.

Genera: 53. *Stipagrostis*
         54. *Aristida*

Tribe XIII. PAPPOPHOREAE

A small tribe closely related to XIV. Eragrostideae but distinguished by its many-lobed lemmas, the lobes either awned or alternating with awns.

Genera: 55. *Enneapogon*
         56. *Schmidtia*

Tribe XIV. ERAGROSTIDEAE

A large tribe of tropical grasses, many of them weeds or pioneers of disturbed places. It is distinguished from V. Poeae by the 3-nerved lemma (except 57. *Aeluropus;* some Poeae 3-nerved) and by other cytological and anatomical characters. Subtribe Monanthochloinae comprises halophytes with many-nerved lemmas; Eleusininae generally have a 3-nerved lemma and several-flowered spikelets; Sporobolinae have a 1-flowered spikelet.

Subtribe MONANTHOCHLOINAE
  Genus: 57. *Aeluropus*

Subtribe ELEUSININAE
  Genera: 58. *Triraphis*
          59. *Leptochloa*
          60. *Halopyrum*
          61. *Trichoneura*
          62. *Dinebra*
          63. *Ochthochloa*
          64. *Eragrostis*
          65. *Coelachyrum*
          66. *Eleusine*
          67. *Acrachne*
          68. *Dactyloctenium*
          69. *Desmostachya*

Subtribe SPOROBOLINAE
  Genera: 70. *Sporobolus*
          71. *Crypsis*

Tribe XV. CYNODONTEAE

A tribe closely related to XIV. Eragrostideae, but differing principally in the possession of only a single fertile floret usually accompanied by 1 or more barren florets. These characters are sometimes unreliable, so the facies of the inflorescence may help, though this is rather hard to define beyond being racemose and never paniculate. Subtribe Chloridinae are the basic stock; Boutelouinae differ by the 3-lobed lemma and Zoysiinae by the reduced deciduous racemes with highly modified spikelets.

Subtribe CHLORIDINAE
  Genera: 72. *Tetrapogon*
          73. *Chloris*
          74. *Enteropogon*
          75. *Cynodon*
          76. *Schoenefeldia*

Subtribe BOUTELOUINAE
  Genus: 77. *Melanocenchris*

Subtribe ZOYSIINAE
  Genera: 78. *Tragus*
          79. *Leptothrium*

Tribe XVI. PANICEAE

A huge tropical tribe in which the spikelets fall with their glumes at maturity. The glumes are small and thin and the spikelet is 2-flowered. The lower floret is male or barren, the upper bisexual. The grain is protected by the toughened upper lemma which may be hard and shiny and resembling a tiny cowrie shell. Subtribe Setariinae has a hard upper lemma; Melinidinae have a cartilaginous upper lemma, paniculate inflorescence and reduced lower glume; Digitariinae also mostly have a cartilaginous upper lemma and reduced lower glume, but the inflorescence usually consists of racemes; Cenchrinae have a highly modified panicle in which sterile branches form a deciduous involucre around the spikelet.

Subtribe SETARIINAE
Genera: 80. *Panicum*
81. *Echinochloa*
82. *Brachiaria*
83. *Paspalum*
84. *Setaria*
85. *Paspalidium*
86. *Stenotaphrum*

Subtribe MELINIDINAE
Genera: 87. *Tricholaena*
88. *Melinis*

Subtribe DIGITARIINAE
Genus: 89. *Digitaria*

Subtribe CENCHRINAE
Genera: 90. *Pennisetum*
91. *Cenchrus*

Tribe XVII. ARUNDINELLEAE

A tropical tribe recognised in Egypt by the tendency of the spikelets to form triads; the 2 florets in the spikelet are dimorphic and are shed from the persistent glumes.

Genus: 92. *Danthoniopsis*

Tribe XVIII. ANDROPOGONEAE

In most cases this tribe can be recognised by its fragile racemes bearing paired spikelets, one of each pair being sessile, the other pedicelled. Usually the spikelets of a pair differ in appearance and sex. Subtribe Saccharinae is regarded as the most primitive since both spikelets of the pair have retained fertility and are similar in appearance. In Sorghinae fertility has been lost from the pedicelled spikelet. In Andropogoninae the inflorescence is reduced to solitary or paired racemes and the sessile floret has a blunt callus inserted into the hollowed internode-tip; Anthistiriinae are similar, but the callus is pointed and applied obliquely to the internode-tip. Rottboellinae are characterised by thickened, swollen raceme-internodes and pedicels. Tripsacinae and Coicinae are monoecious, the latter protecting the disseminule with a hard flask-like spatheole.

7

Subtribe SACCHARINAE
  Genera: 93. *Saccharum*
          94. *Miscanthus*
          95. *Imperata*
          96. *Pogonatherum*

Subtribe SORGHINAE
  Genera: 97. *Sorghum*
          98. *Vetiveria*
          99. *Chrysopogon*
         100. *Dichanthium*

Subtribe ANDROPOGONINAE
  Genera: 101. *Andropogon*
         102. *Cymbopogon*

Subtribe ANTHISTIRIINAE
  Genera: 103. *Hyparrhenia*
         104. *Themeda*

Subtribe ROTTBOELLIINAE
  Genera: 105. *Elionurus*
         106. *Hemarthria*
         107. *Lasiurus*

Subtribe TRIPSACINAE
  Genus: 108. *Zea*

Subtribe COICINAE
  Genus: 109. *Coix*

KEY TO TRIBES

1. Creeping perennial with the spikelets enclosed in the uppermost sheath, visible only by virtue of the long protruding filaments and stigmas (*Pennisetum clandestinum*) . . . . XVI. Paniceae (p.46)
+ Not as above; some spikelets always exserted 2

2. Spikelet urn-like, comprising 2 opposing lemmas fused into a tube, the 2 paleas fused into a septum within the tube, and the whole embraced below by an inflated sheath; glumes absent . . . . III. Lygeae (p.11)
+ Spikelet not as above, lemmas not fused into a tube 3

3. Spikelets 1- to many-flowered, breaking up above the more or less persistent glumes, or if falling entire then not 2-flowered with the lower floret male or barren and the upper bisexual, nor dorsally compressed; if 2-flowered and falling entire then strongly laterally compressed (26. *Holcus*); plants not monoecious (50. *Cortaderia* gynodioecious with plumose panicle) 4
+ Spikelets 2-flowered, falling entire at maturity, with the upper floret bisexual and the lower male or barren and in the latter case often much reduced; spikelets usually dorsally compressed 22

4. Inflorescence a fragile cylindrical raceme with the spikelets sunk in hollows in the axis; glumes collateral, concealing the hollows . . . . VI. Hainardieae (p.18)
+ Inflorescence not a fragile cylindrical raceme 5

5. Ovary with a fleshy hairy apical appendage, the styles arising from beneath it 6
+ Ovary without a fleshy hairy apical appendage; styles clearly terminal 7

6. Inflorescence a spicate raceme (or a false raceme with up to 3 spikelets at a node) . . . . X. Triticeae (p.26)
+ Inflorescence a panicle (sometimes reduced to a loose raceme in depauperate specimens) . . . . IX. Bromeae (p.24)

7. Lemmas deeply cleft into 9 awned lobes or 6 lobes alternating with 5 awns . . . . XIII. Pappophoreae (p.36)
+ Lemmas entire or 2-lobed, awnless or 1- to 3-awned 8

8. Spikelets containing 1 fertile floret (except 72. *Tetrapogon*), with or without 1 or 2 male or barren florets below it or 1 or more above . . 9
+ Spikelets containing 2 or more fertile florets 18

9. Glumes absent; palea 1-keeled . . . . I. Oryzeae (p.11)
+ Glumes, or at least one of them, well developed; palea 2-keeled . . 10

10. Spikelets arranged in 1-sided racemes or in short deciduous racemelets arranged along an axis (and the spikelets highly modified) . . . . XV. Cynodonteae (p.44)
+ Spikelets in panicles, these either open or contracted and spike-like 11

11. Spikelets strictly 1-flowered 12
+ Spikelets 2- to 3-flowered, or with a clavate mass of several reduced sterile lemmas above the fertile 15

12. Lemma with 3 awns, these often connate below into a twisted column
   . . . . XII. Aristideae (p.34)
 + Lemma awnless or with a single awn 13

13. Lemma indurated at maturity, awned from the tip . . IV. Stipeae (p.11)
 + Lemma hyaline or membranous at maturity 14

14. Glumes longer and firmer than the hyaline lemma; lemma sometimes
   awned; grain with adherent pericarp . . . . . .VIII. Aveneae (p.18)
 + Glumes and lemmas similar in texture, the former often the shorter;
   lemmas always awnless; grain with loose pericarp which on wetting
   swells and ejects the seed . . . . XIV. Eragrostideae (p.38)

15. Spikelets with a clavate mass of several reduced sterile lemmas above
   the fertile . . . . VII. Meliceae (p.18)
 + Spikelets with the sterile lemmas below the fertile 16

16. Florets 2, the lower well developed but male or barren
   . . . . XVII. Arundinelleae (p.52)
 + Florets 3, or if only 2 then the lower small, chaffy and vestigial . . .17

17. Sterile lemmas small, chaffy and vestigial, the fertile hardened and
   shiny, shorter than the glumes . . . . .VIII. Aveneae (p.18)
 + Sterile lemmas longer than the fertile, at least one of them transversely
   ridged or wrinkled, the florets exceeding the glumes II. Ehrharteae (p.11)

18. Tall reed-like or tussock-forming grasses with large plumose panicles
   . . . . XI. Arundineae (p.29)
 + Slender grasses; panicles not plumose 19

19. Glumes shorter than the lowest lemma, with the florets distinctly
   exserted; lemmas awnless or with a straight awn from the entire or
   2-lobed tip 20
 + Glumes longer than the lowest lemma, often as long as the spikelet and
   enclosing the florets 21

20. Lemmas 1- to 3-nerved (9- to 11-nerved in 57. *Aeluropus*); spikelets
   all fertile . . . . XIV. Eragrostideae (p.38)
 + Lemmas 5-nerved, if 3-nerved then inflorescence a dichotomously
   branched panicle (19. *Cutandia*) or lemma clavate-hairy on the back
   (17. *Desmazeria*); sterile and fertile spikelets sometimes intermixed in
   the panicle (10. *Cynosurus*, 11. *Lamarckia*) . . . . . .V. Poeae (p.13)

21. Ligule a membrane; lemma awned from the back, rarely awnless
   . . . . VIII. Aveneae (p.18)
 + Ligule a line of hairs; lemma awned from the sinus of the prominently
   2-lobed tip, rarely awnless . . . . XI. Arundineae (p.29)

22. Spikelets solitary, rarely paired but then those of a pair alike; glumes
   membranous, the lower mostly the smaller and sometimes suppressed;
   upper lemma indurated, cartilaginous to crustaceous, usually awnless
   but sometimes mucronate . . . . XVI. Paniceae (p.46)
 + Spikelets typically paired with 1 sessile and the other pedicelled, those
   of a pair usually dissimilar, rarely with the spikelets all alike; glumes as
   long as the spikelet and enclosing the florets, more or less rigid and
   firmer than the hyaline or membranous lemmas; upper lemma often
   with a geniculate awn; rarely the plants monoecious (108. *Zea*, 109.
   Coix) . . . . XVIII. Andropogoneae (p.52)

ENUMERATION OF TRIBES, GENERA AND SPECIES

I. ORYZEAE

Spikelets with 2 sterile lemmas below the fertile . . . . **1. *Oryza***
Spikelets strictly 1-flowered . . . . **2. *Leersia***

1. *Oryza*

Spikelets persistent; ligule of lower leaves acute, 15–45mm long. Cultivated cereal (rice) grown in wet fields. Mma Mp Nv O. *Cult. in warm temp. regions* . . . . ***O. sativa***

2. *Leersia*

Rhizomatous perennial; spikelets deciduous; ligule 1–2mm long. Wet ditches, canals and rice-fields. (Mma) Mp Nv O. *Throughout the tropics* . . . . ***L. hexandra***

II. EHRHARTEAE

3. *Ehrharta*

Tufted perennial. Introduced to Ras el Hekma in the early 1950s but probably no longer there. *Namibia and South Africa* . . . . . . ***E. calycina***

III. LYGEAE

4. *Lygeum*

Tufted perennial with wiry leaves. Rocky places, sandy, often calcareous, plains, and saltmarshes. Mma (Mp). *Mediterranean region* . . . ***L. spartum***

IV. STIPEAE

Callus pungent, densely bearded; spikelets laterally compressed or terete . . . . **5. *Stipa***
Callus obtuse, scarcely distinct from the lemma, glabrous; spikelets dorsally compressed . . . . **6. *Oryzopsis***

5. *Stipa* (Plate 1)

1. Annual; awns eventually twisted together to form a tail at the summit of the panicle. Desert and coastal sands and rocky slopes. Mma Mp Nv Dl Dg (Da) Di S R Sa. *Mediterranean region to NW India; South Africa*
. . . . ***S. capensis***
+ Perennials; awns not twisted together into a tail, but sometimes entangled 2

11

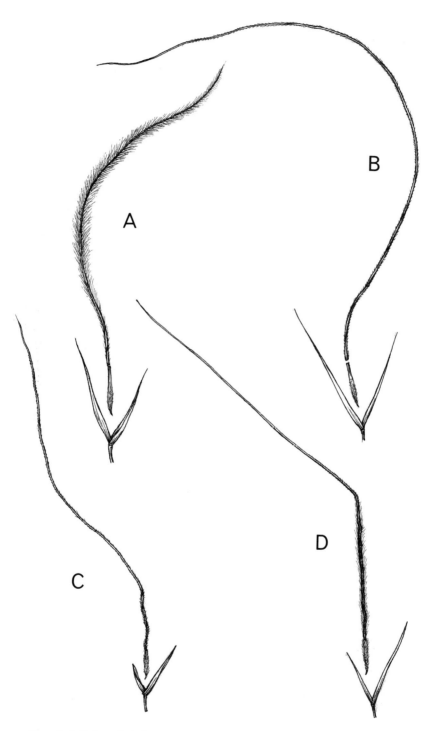

Plate 1. Spikelets of: A, *Stipa arabica*; B, *S. lagascae*; C, *S. parviflora*; D, *S. capensis*. Not to scale.

2. Awn plumose to the naked eye for its whole length. Stony soils. S. *E mediterranean region to Pakistan* . . . . *S. arabica*
+ Awn scabrid or shortly pubescent for its whole length, if the latter then the hairs barely discernible to the naked eye 3

3. Awn 20–25cm long, shortly pubescent. Sand. Mma (Nv Dl Di S). *Spain and Sicily and along the shores of the southern Mediterranean* (See Appendix 1 note 1) . . . . *S. lagascae*
+ Awn 6–10(13)cm long, scabrid. Sandy and stony soils. Mma Mp Dg Di (O) S. *Mediterranean region to Iran* (See Appendix 1 note 2) *S. parviflora*

6. *Oryzopsis*

Lemma glabrous except for 3 basal tufts of hair; panicle-branches in whorls. Loosely tufted perennial arising from a thick rootstock, up to 120cm high. Damp sandy soils in gardens and orchards. Mma (Mp) Nv (Dg) Di S. *Mediterranean region and SW Asia* . . . . *O. miliacea*
Lemma hairy; panicle-branches not whorled. Densely tufted perennial up to 100cm high. Stony soils. S. *Mediterranean region and SW Asia*
. . . . *O. holciformis*

V. POEAE

1. Fertile spikelets intermixed with sterile spikelets 2
+ Fertile spikelets not accompanied by sterile spikelets 3

2. Fertile spikelet with 1 sterile companion, the latter persistent
. . . . **10. *Cynosurus***
+ Fertile spikelet with several sterile companions, falling together
. . . . **11. *Lamarckia***

3. Inflorescence a capitate panicle . . . . **20. *Ammochloa***
+ Inflorescence an open or contracted panicle or an elongated raceme 4

4. Inflorescence an elongated bilateral raceme, the spikelets edgeways on to the axis; lower glume absent (except in the terminal spikelet)
. . . . **8. *Lolium***
| Inflorescence a panicle, or if a raceme then both glumes present . . 5

5. Plants perennial 6
+ Plants annual 8

6. Lemmas rounded on the back . . . . **7. *Festuca***
+ Lemmas keeled throughout 7

7. Lemma subacute, not spinulose on the keel, or spikelets viviparous; panicle contracted but neither lobed nor secund . . . . . . **13. *Poa***
+ Lemma acuminate, spinulose on the keel; panicle secund, lobed
. . . . **14. *Dactylis***

8. Lemma awned, pedicels not inflated . . . . **9. *Vulpia***
+ Lemma awnless, if shortly awned then pedicels inflated 9

9. Lemma 5- to 11-nerved, if apparently 3-nerved then clavate-hairy below
. 10
+ Lemma 3-nerved, not clavate-hairy 14

10. Lemma rounded on the back . . . . **18.** *Catapodium*
+ Lemma keeled throughout 11

11. Pedicels stout; inflorescence a 1-sided panicle with short stiff branches, or a raceme . . . . **17.** *Desmazeria*
+ Pedicels slender; inflorescence a panicle, open or contracted but not 1-sided 12

12. Lemma orbicular to oblate . . . . **12.** *Briza*
+ Lemma lanceolate to ovate 13

13. Panicle-branches not whorled; lemmas narrowly ovate in profile **13.** *Poa*
+ Panicle-branches whorled; lemma lanceolate to narrowly oblong in profile . . . . **15.** *Eremopoa*

14. Panicle-branches and pedicels slender, flexuous, persistent **16.** *Sphenopus*
+ Panicle-branches and pedicels stout, stiff, deciduous . . . **19.** *Cutandia*

7. *Festuca*

Leaves short, stiff and pungent, up to 1mm across. Tufted perennial with fibrous base. Almost certainly recorded in error. *Turkey and Lebanon*
. . . . ***F. pinifolia***
Leaves long and flat, 5–10mm across. Loosely tufted perennial without fibrous base. Introduced to Ras el Hekma in the early 1950s but probably no longer there. *Temperate Eurasia* . . . . ***F. elatior***

8. *Lolium* (Plate 2, A–D)

1. Lemma elliptical to ovate, very turgid at maturity especially towards the base; mature fruit not more than 3 times as long as wide. Weed of cereals. Mma Md Mp Nv Nn (Di) O (S). *Mediterranean region and SW Asia* . . . . ***L. temulentum***
+ Lemma oblong to lanceolate, not turgid at maturity; mature fruit more than 3 times as long as wide 2

2. Perennial with non-flowering shoots at anthesis; spikelets 2- to 10-flowered (rarely up to 14-flowered), usually awnless; leaves flat or folded when young. Mostly in areas of cultivation. Mma (Mp) Nv (Dg Da Di) O (S). *Europe, N Africa and temp. Asia* . . . . . . ***L. perenne***
+ Annual without non-flowering shoots at anthesis, or if biennial or perennial then spikelets 11- to 22-flowered; leaves rolled when young 3

3. Spikelets 11- to 22-flowered, rarely less; glume always less than half as long as the spikelet; annual, biennial or short-lived perennial. Weed of cereals. (Mma) Mp Nv (Dl) Di O R. *Central and southern Europe, NW Africa and SW Asia* . . . . ***L. multiflorum***
+ Spikelets 3- to 11-flowered; glume usually more than half as long as the spikelet. Cultivated fields and gardens. Mma Md Mp Nv (Di) O S. *Southern Europe and the Mediterranean region to C Asia* . . . ***L. rigidum***

9. *Vulpia*

1. Spikelets pectinate. Sand overlying limestone. Mma (Mp) Di (S). *N Africa, Syria and Palestine* . . . . ***V. pectinella***
+ Spikelets more or less appressed to the branches 2

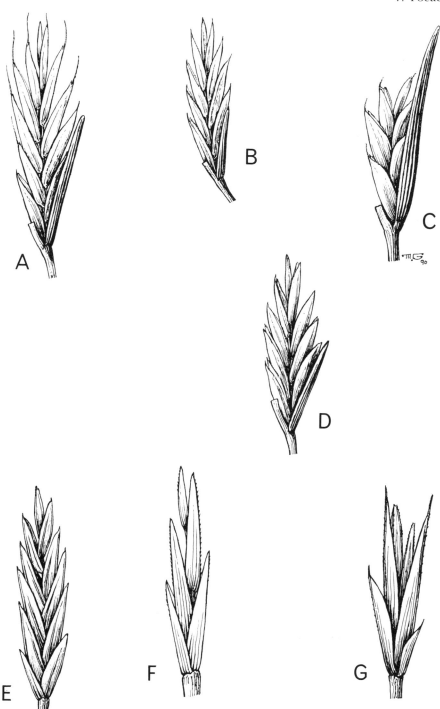

Plate 2. Spikelets of: A, *Lolium multiflorum*; B, *L. perenne*; C, *L. temulentum*; D, *L. rigidum*; E, *Cutandia maritima*; F, *C. dichotoma*; G, *C. memphitica*. Not to scale.

15

2. Spikelets deciduous in triads. Margins of sandy fields and thin soils overlying limestone. Mma Md Mp Nv (Di). *N Africa* . . . . . ***V. inops***
+ Spikelets solitary, not falling in triads 3

3. Lemma with pointed callus. Sand. Mma Md Mp Nv. *W Europe and the Mediterranean region eastwards to the Caucasus* . . . . . ***V. fasciculata***
+ Lemma with rounded callus 4

4. Lemma 1.3–1.9mm wide; lower glume 2.5–5mm long, ½–¾ as long as the upper. Unconfirmed. (Mma S). *W & C Europe and the Mediterranean region; mtns. of trop. Africa* . . . . ***V. bromoides***
+ Lemma 0.8–1.3mm wide; lower glume 0.4–2.5mm long, ¹/₁₀–²/₅ as long as the upper. Sand. Mma S. *Europe and temp. Asia* . . . . ***V. myuros***

10. *Cynosurus*

Awns of sterile spikelets 15–20mm long, pink at the base; uppermost sheath often reaching or embracing the base of the panicle at anthesis. Calcareous sand. Mma (Mp). *E Mediterranean region* . . . . . ***C. coloratus***
Awns of sterile spikelets 6–15mm long, usually pallid; uppermost sheath far below the panicle at anthesis. Calcareous sand. Mma. *Mediterranean region to C Asia* . . . . ***C. echinatus***

11. *Lamarckia*

Small annual with spike-like golden yellow panicle tinged with purple; spikelets dimorphic, falling in clusters. Sandy and stony soils and areas of cultivation. Mma Nv Di (S). *Macaronesia to NE Africa and C Asia* . . ***L. aurea***

12. *Briza*

Spikelets 3–5mm long, numerous in the panicle. Weed of cultivation. (Mma) Md Nv. *Mediterranean region* . . . . ***B. minor***
Spikelets 14–25mm long, up to 12 in the panicle. An introduced decorative plant. *Mediterranean region* . . . . ***B. maxima***

13. *Poa*

1. Plants perennial with bulbous base. Rarely producing proliferating spikelets. Mountains and rocky deserts. Dg Di S. *SW Asia to northern India* . . . . ***P. sinaica***
+ Plants annual .2

2. Lower panicle-branches erecto-patent after anthesis; spikelets with rather distant florets; anthers 0.2–0.5mm long, scarcely longer than wide. Weed of fields and gardens. Mma Nv Di. *Southern Europe to northern India and C Asia; intr. in S America* . . . . . . ***P. infirma***
+ Lower panicle-branches spreading or deflexed after anthesis; spikelets with crowded florets; anthers 0.6–0.8(1)mm long, 2–3 times as long as wide. Weed of fields and gardens. Mma Mp Nv Di O. *Cosmopolitan* . . . . ***P. annua***

14. *Dactylis*

Coarsely tufted perennial with the vegetative shoots characteristically strongly laterally compressed, and the spikelets in dense, 1-sided clumps. Rocky ground; introduced in areas of cultivation. Mma (Nv*). *Europe and temp. Asia* (See Appendix 1 note 3) . . . . **D. glomerata**

15. *Eremopoa*

Delicate annual with whorled panicle-branches. Shady places among rocks. S. *SW & C Asia to Afghanistan* . . . . **E. altaica**

16. *Sphenopus*

Delicate annual with slender panicle-branches and pedicels, the latter inflated at the tip; lower glume minute, the upper only slightly longer; the whole plant often suffused with purple. Sand and alluvium. Mma Md Mp Nv (Di) O. *Mediterranean region to C Asia* . . . . **S. divaricatus**

17. *Desmazeria*

Annual with stiff panicle branched below, racemose above. Coastal sand and among calcareous rocks. Mma. *N Africa and Palestine*
. . . . **D. philistaea** subsp. **rohlfsiana**

18. *Catapodium*

Annual with a stiff, rather narrow panicle with short rigid 3-angled branches and pedicels. Coastal sand. Mma. *Southern and western Europe, Mediterranean region and SW Asia* . . . . **C. rigidum**

19. *Cutandia* (Plate 2, E–G)

1. Glumes 3- to 5-nerved; spikelets 5- to 12-flowered, ovate. Sand, mainly on the coast. Mma Di. *Mediterranean region* . . . . . . **C. maritima**
+ Glumes 1-nerved; spikelets 3- to 18-flowered, oblong-linear 2

2. Lemma-tip produced into a short awn, or at least mucronate; lemma 7.5–8.5mm long. Sandy, sometimes saline soils. Mma Md Mp Nv Dl (Dg Da) Di O S (R) Sa. *Mediterranean region and SW Asia* . . **C. memphitica**
+ Lemma-tip acute or at most slightly mucronate; lemma 4.5–5.5mm long. Sand. Mma Md Mp Nv Di. *N Africa and SW Asia* . . **C. dichotoma**

20. *Ammochloa*

Annual; panicle capitate, nestling among the basal sheaths, giving the plant the aspect of some species of *Cyperus*. In sand and among calcareous rocks. Mma (Md) Mp Nv (Di). *Mediterranean region and SW Asia* **A. palaestina**

VI. HAINARDIEAE

21. *Parapholis*

1. Keel of glumes wingless. Inflorescence strongly curved. Sand. Mma
   Md Mp Nv Di O. *Western Europe, Mediterranean region and SW Asia*
   . . . . *P. incurva*
+ Keel of glumes distinctly winged 2

2. Anthers less than 1mm long; plant usually c.10cm high, stout, with
   clustered racemes and usually straight. Sandy and stony soils, but
   much confused with other species in the genus and probably under-
   recorded. Mma (Md Mp Nv Dl Dg Di O S). *E Mediterranean region*
   . . . . *P. marginata*
+ Anthers more than 2mm long; plant usually more than 10cm high,
   slender, with solitary racemes, straight or nearly so. Possibly recorded
   in error. (Mma Md Mp Nv Di O S). *Mediterranean region* . . *P. filiformis*

VII. MELICEAE

22. *Melica*

Densely tufted rhizomatous perennial; fertile lemma densely pilose.
Stony soils. S. *E Mediterranean region and SW Asia to Pakistan* . . . *M. persica*

VIII. AVENEAE

1. Spikelets with 2 or more fertile florets, sometimes the uppermost
   reduced 2
+ Spikelets with 1 fertile floret, sometimes with 1 or 2 sterile florets below
   it reduced to small chaffy scales; if 2-flowered with the florets more or
   less equal in size but the lower fertile and the upper male, then spikelet
   falling entire 5

2. Spikelets large, 16–50mm long, pendant in a large open panicle up to
   40cm long . . . . **23. *Avena***
+ Spikelets not more than 7.5mm long 3

3. Awn with a ring of hairs at the junction of the column and the clavate
   limb . . . . **27. *Corynephorus***
+ Awn without a ring of hairs or quite absent 4

4. Lemmas 2-toothed, the teeth often aristulate, and sometimes also with
   a dorsal awn . . . . **24. *Trisetaria***
+ Lemmas obtuse to acute, awnless or with a short awn-point from just
   below the tip . . . . **25. *Rostraria***

5. Spikelets 2-flowered, the florets more or less equal in size, falling
   entire . . . . **26. *Holcus***
+ Spikelets breaking up above the persistent glumes, or if falling entire
   then strictly 1-flowered 6

6. Spikelets 2- to 3-flowered, the uppermost floret fertile, hard and
   shining, the 1 or 2 sterile florets below it reduced to small chaffy
   scales . . . . **28. *Phalaris***

+ Spikelets strictly 1-flowered, the floret not hard and shining (if so, see
   IV. Stipeae) 7

7. Spikelets breaking up above the persistent glumes 8
+ Spikelets falling entire 12

8. Plants perennial 9
+ Plants annual .10

9. Lemma hyaline, not compressed, less than 3mm long . . . **29. Agrostis**
+ Lemma thinly coriaceous, strongly laterally compressed and keeled,
   8–12mm long . . . . **30. Ammophila**

10. Inflorescence an ovoid, softly and densely hairy head . . . **33. Lagurus**
+ Inflorescence a spike-like panicle, but not hairy 11

11. Glumes swollen and slightly toughened below; lemma 1-awned or
   awnless . . . . **32. Gastridium**
+ Glumes neither toughened nor swollen below; lemma 3-awned
   . . . . **31. Triplachne**

12. Spikelets shed with a basal stipe . . . . **34. Polypogon**
+ Spikelets shed without a basal stipe 13

13. Lemma dorsally awned . . . . **35. Alopecurus**
+ Lemma awnless . . . . **36. Phleum**

23. *Avena*

1. Rhachilla continuous between the florets, not disarticulating at
   maturity 2
+ Rhachilla articulated between the florets, these falling separately at
   maturity 4

2. Rhachilla not disarticulating above the glumes at maturity (but
   fracturing unevenly under pressure); glumes 20–25mm long.
   Cultivated cereal (oats) sometimes also found an as escape. Mma Md
   Mp Nv Di O. *Cult. in temp. regions* . . . . **A. sativa**
+ Rhachilla disarticulating above the glumes at maturity (and producing
   a smooth, obliquely elliptical scar at the base of the lowest lemma);
   glumes 25–50mm long; wild oat (*A. sterilis*) 3

3. Glumes 30–50mm long; lowest lemma 25–40mm long. Weed of cereals.
   Mma (Mp) Nv (Nn) Di (O) S. *Mediterranean region to C Asia*
   . . . . **A. sterilis** subsp. **sterilis**
+ Glumes 25–30mm long; lowest lemma 20–25mm long. Weed of cereals
   and vegetables. Mma Mp Nv (Nn) Di O S. *Mediterranean region to India*
   . . . . **A. sterilis** subsp. **ludoviciana**

4. Tip of lemma with 2 acute teeth or these at most shortly mucronate.
   Weed of cultivation. Mma Md Mp Nv Nn Dl (Dg Di) O S. *Temp. Old World*
   . . . . **A. fatua**
+ Tip of lemma with 2 awned teeth, the awns up to 12mm long 5

5. Callus long and pungent, the scar linear. Near the coast. Mma (Nv).
   Mediterranean region except the north . . . . **A. longiglumis**
+ Callus short with oval scar (*A. barbata*) 6

6. Spikelets 20–30mm long; lowest lemma 16–20mm long. Weed of
   cultivation. Mma Md Mp (Nv Dl Dg Da Di O) S. *Europe and N Africa to*
   C Asia . . . . **A. barbata** subsp. **barbata**

+ Spikelets 16–20mm long; lowest lemma 12–16mm long. Weed of cereals. Mma (Mp) Nv Dl Dg (Da) Di (O) S. *Mediterranean region and SW Asia* . . . . **A. barbata** subsp. *wiestii*

24. Trisetaria

1. Panicle loose and interrupted. Sandy and rocky ground. Mma (Nv Da Di). *Palestine, Libya and Spain* . . . . **T. macrochaeta**
+ Panicle dense, spike-like 2

2. Central awn of lemma inserted at or below the middle. In sandy soil and amongst limestone rocks. Mma (Mp) Di. *Syria and Palestine* . . . . **T. glumacea**
+ Central awn of lemma inserted in the upper ¼, or sometimes quite absent 3

3. Anthers 0.5–1mm long. Mobile dunes. Mma Md Mp. *N Africa and SW Asia* . . . . **T. linearis**
+ Anthers 1.5–2mm long. Mobile dunes. Mp. *Syria, Lebanon and Palestine* . . . . **T. koelerioides**

25. Rostraria (Plate 3)

1. Glumes subequal, the lower minutely longer than the upper, the lower often (the upper rarely) densely woolly. Alluvial and cultivated soils. Mma Mp Nv Dl Dg Di S (R) Sa. *Mediterranean region to NW India* . . . . **R. pumila**
+ Glume unequal, the lower shorter and narrower than the upper . . . 2

2. Lemma very obtuse. Sand. Mp. *Turkey, Cyprus and Syria* **R. obtusiflora**
+ Lemma subacute 3

3. Panicle ovoid, bristly; lemma with stiff awn 3–5mm long. Unconfirmed. (Mma). *Mediterranean region* . . . . **R. hispida**
+ Panicle oblong, not bristly; lemma with slender awn usually less than 3mm long 4

4. Lemma with terminal awn. Desert sand and cultivated soils. Mma Md Mp Nv Nn (Dl Dg) Di O S. *Mediterranean region to NW India* **R. cristata**
+ Lemma with dorsal, or at least clearly subterminal awn. Weed of cultivated fields and palm groves. O. *Saharan region* . . . **R. rohlfsii**

26. Holcus

Annual; spikelets 2-flowered, falling entire. Sand; not native. Mma* Nv*. *Mediterranean region* . . . . **H. annuus**

27. Corynephorus

Annual; awn of unique structure comprising a twisted column articulated with a clavate limb, and a ring of short hairs at the joint. Coastal sand. Mma Md Mp. *Mediterranean region to the Caspian* . . . **C. divaricatus**

28. Phalaris (Plate 4)

1. Spikelets falling in clusters at maturity, one (rarely two) of the spikelets fertile and the remainder male or sterile, forming an involucre (if spikelets falling singly then fertile lemma quite glabrous) 2

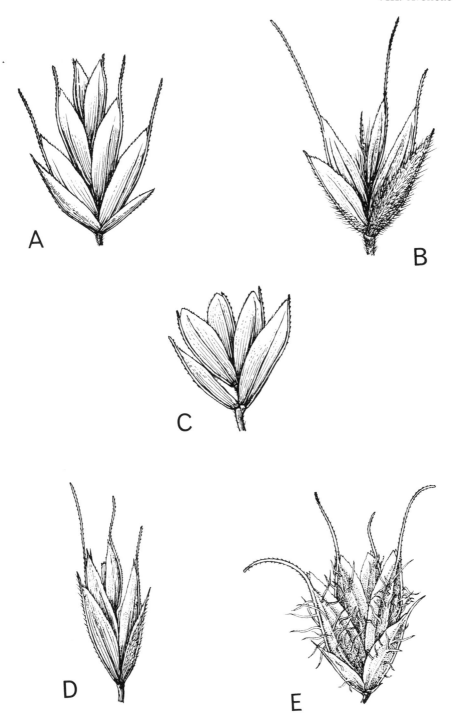

Plate 3. Spikelets of: A, *Rostraria cristata*; B, *R. pumila*; C, *R. obtusiflora*; D, *R. rohlfsii*; E, *R. hispida*. Not to scale.

21

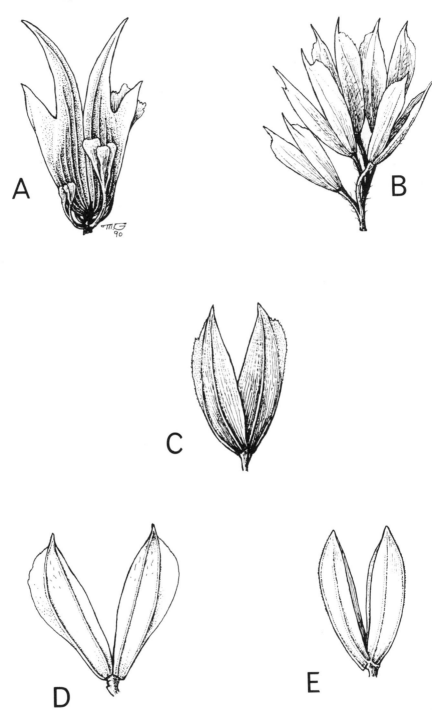

Plate 4. Spikelets of: A, *Phalaris paradoxa*; B, *P. coerulescens*; C, *P. minor*; D, *P. canariensis*; E, *P. aquatica*. Not to scale.

+ Spikelets not falling in clusters, the florets deciduous from the persistent glumes, all fertile; fertile lemma sparsely to densely pubescent 3

2. Annual. Canal banks, cultivated fields and waste ground. Mma Mp Nv Dl (Dg) Di O. *Southern Europe, Mediterranean region and SW Asia* . . . . **P. paradoxa**
+ Perennial with tuberous base. Damp ground, usually in wadis. Di. *Mediterranean region* . . . . **P. coerulescens**

3. Annuals 4
+ Perennials 5

4. Sterile floret 1, either well developed (more than 0.5mm long) or obsolete (less than 0.3mm long); wing of glumes toothed or erose. Weed of roadsides and cultivated areas. Mma Md Mp Nv (Nn) Dl (Dg Da) Di O S. *Mediterranean region to NW India* . . . . **P. minor**
+ Sterile florets 2, broad and chaffy, more than half as long as the fertile lemma; wing of glumes entire. Introduced in bird-seed. Mma* Nv*. *Mediterranean region* . . . . **P. canariensis**

5. Perennial with long creeping rhizomes, lobed panicle and variegated leaves. An escape from gardens, unknown as a wild plant . . . . **P. arundinacea** var. **picta**
+ Tufted perennial with short rhizomes and bulbous swellings at the base. Introduced to Ras el Hekma in the early 1950s but probably no longer there. *Mediterranean region* . . . . **P. aquatica**

29. *Agrostis*

Stoloniferous perennial. Damp soils. Di (S). *Europe and temp. Asia* . . . . **A. stolonifera**

30. *Ammophila*

Tall, deeply rhizomatous perennial with rigid, tightly inrolled leaves. A useful sand-binder. Coastal sand. Mma Md Mp. *Coasts of western Europe* . . . . **A. arenaria** subsp. **arundinacea**

31. *Triplachne*

Annual; panicle dense, shortly cylindrical to ovoid; lemma 3-awned. Sandy fields. Mma Mp. *Mediterranean region* . . . . **T. nitens**

32. *Gastridium*

Annual; glumes swollen and slightly toughened below. Mma. *Mediterranean region and E trop. Africa* . . . . **G. phleoides**

33. *Lagurus*

Annual with ovoid, densely fluffy white panicle. Sand, especially near the coast. Mma (Mp). *Mediterranean region* . . . . **L. ovatus**

34. *Polypogon*

1. Perennial; glumes awnless. Cultivated and damp alluvial soils. Mma (Mp) Nv Nn (Dl Dg Da) Di O S (R). *Europe and the Mediterranean region to C Asia* . . . . **P. viridis**
+ Annual; glumes awned 2

2. Lemma awned; margins of glumes slightly ciliate above. Cultivated areas and damp soils along canal-banks. Mma Mp Nv Nn (Dl) Dg Da Di O S R. *Europe, Mediterranean region and temp. Asia* . . . **P. monspeliensis**
+ Lemma awnless; margins of glumes densely ciliate above. Alluvial soils and coastal sand. Mma Md (Mp) Nv Nn S. *Europe, Mediterranean region and temp. Asia* . . . . **P. maritimus**

35. *Alopecurus*

Annual; spikelets often dark-coloured in a narrow, spike-like panicle; glumes connate below. Cultivated ground. Mma Nv Di. *Europe and temp. Asia* . . . . **A. myosuroides**

36. *Phleum*

Annual. Unconfirmed in Egypt. (Mp). *W & C Europe and the Mediterranean region eastwards to Pakistan* . . . . **P. subulatum**
Perennial. Said to have been introduced along canal-banks, but no extant material available. (Nv*). *Europe; widely introduced in temp. countries* . . . . **P. pratense**

IX. BROMEAE

Lemma 1-awned; panicle open or contracted . . . . **37. Bromus**
Lemma 9-awned; panicle very dense, very tightly contracted
. . . . **38. Boissiera**

37. *Bromus* (see Appendix 1 note 4)

1. Lemmas strongly laterally compressed and keeled (Sect. *Ceratochloa*). Cultivated ground. Mma* Mp* Nv* Di* O*. *Introduced from the New World* . . . . **B. catharticus**
+ Lemmas rounded on the back 2

2. Spikelets ovate, lanceolate, elliptic or oblong, tapering towards the tip (except at anthesis); lower glume 3- to 7-nerved; upper glume 5- to 9-nerved (Sect. *Bromus*) 3
+ Spikelets oblong or cuneate, gaping at maturity; lower glume 1-nerved; upper glume 3-nerved 9

3. Awn terete or triquetrous, inserted less than 1.5mm from the lemma-tip, straight or weakly divaricate in fruit. Annual, often with ciliolate glume-margins. Cultivated ground and along water-courses. (Mma) Nv (Di). *SW Asia* . . . . **B. aegyptiacus**
+ Awn distinctly flattened at the base, inserted more than 1.5mm from the lemma-tip, often patent or deflexed in fruit 4

4. Panicle dense, erect, the branches and pedicels much shorter than the spikelets 5

+ Panicle lax, erect or nodding, at least some branches and pedicels as
  long as or longer than the spikelets 7

5. Spikelets 25–50mm long (excluding awns); lemmas 12–15mm long;
   awns 12–15mm long, usually twisted below and patent. Unconfirmed
   in Egypt. (Nv). *E Mediterranean region* . . . . **B. alopecuros**
+ Spikelets 12–20mm long (excluding awns); lemmas 6.5–9mm long;
   awns 6.5–9(11) mm long, usually only slightly twisted below 6

6. Lemma not more than 2mm wide; panicle ovoid-cuneate or
   verticillate. Cultivated ground and wadi-beds. Mma Md Mp Nv Di.
   Mediterranean region to NW India . . . . **B. scoparius**
+ Lemma 4–5mm wide; panicle ovoid to ovoid-oblong, sometimes with
   few spikelets. Roadside ditches and waste-ground. Nv. *Endemic to Egypt*
   . . . . **B. javorkae**

7. Panicle lax, erect, the pedicels mostly equalling or shorter than the
   spikelets. Nv. *Europe and the Mediterranean region eastwards to C Asia*
   . . . . **B. lanceolatus**
+ Panicle lax, nodding, the pedicels mostly longer than the spikelets   8

8. Lemmas horny, markedly rhombic with inconspicuous nerves and
   blunt apical teeth; margins of lemma overlapping those of the
   adjacent lemma; awns usually strongly reflexed in fruit. Sand. Mma
   (Dg S). *C Europe eastwards to Japan; widely introduced* . . . **B. japonicus**
+ Lemmas papery, evenly curved along the margins, with conspicuous
   nerves and usually acuminate apical teeth; margins of lemma
   somewhat inrolled at maturity and lemmas slightly divaricate; awns
   usually straight or weakly divaricate in fruit. Sandy and rocky ground.
   Di S R Sa. *Trop. & southern Africa to Europe and E Asia* . . . **B. pectinatus**

9. Perennial (Sect. *Pnigma*). Lemmas awnless or almost so. Agricultural
   weed. Nv*. *Europe and temp. Asia* . . . . **B. inermis**
+ Annuals (Sect. *Genea*) 10

10. Lemma more than 20mm long. Cultivated fields. Mma Md Mp Nv Nn
    Di O (S). *C & S Europe, N Africa and SW Asia* . . . . . . **B. diandrus**
+ Lemma less than 20mm long 11

11. Panicle drooping, very lax, most branches and pedicels as long as or
    longer than the spikelets 12
+ Panicle erect, often dense, most branches and pedicels shorter than
    the spikelets or spikelets subsessile 13

12. Panicle simple, the branches each bearing only 1 or 2 spikelets; lemma
    more than 12mm long. Unconfirmed in Egypt. *Europe and SW Asia*
    . . . . **B. sterilis**
+ Panicle always compound, usually secund with the longer branches
    bearing at least 4 spikelets each; lemma 10–12(14) mm long. Sand.
    Mma Md Mp (Dg) Di S. *Temp. Old World* . . . . **B. tectorum**

13. Panicle lax, the spikelets not densely crowded; panicle-branches
    10mm long or more; lemma at least 3mm wide. Sandy soils and
    cultivated ground. Mma Mp (Dg Di) S. *Mediterranean region and SW
    Asia* . . . . **B. madritensis**
+ Panicle very dense, the spikelets crowded on branches 1–10mm long;
    lemma less than 3mm wide 14

14. Lemma 1–1.5mm wide; panicle up to 4cm long with 1–2 spikelets on each branch; awns somewhat curved outwards in fruit. Sandy and stony soils. Mma Md Mp Nv (Dl Dg) Di (S) R Sa. *Mediterranean region and SW Asia* . . . . **B. fasciculatus**
+ Lemma c.2mm wide; panicle usually more than 4cm long, the branches each with 4–5 spikelets; awns more or less straight in fruit. Sandy soils and as a weed of cultivation. Mma Md Mp Nv (Dl Di) O (S). *Mediterranean region and SW Asia* . . . . **B. rubens**

38. *Boissiera*

Small annual with dense ovoid panicle and 9-awned lemmas. Sandy and stony deserts. Nv Dg S. *Mediterranean region to C Asia* . . . . . **B. squarrosa**

X. TRITICEAE

1. Plants perennial 2
+ Plants annual 3

2. Glumes keeled to the base; spikelets pectinate . . . . . **44. Agropyron**
+ Glumes not keeled; spikelets not pectinate . . . . **40. Elymus**

3. Spikelets in groups of 2 or 3 at each node of the raceme 4
+ Spikelets borne singly at each node 6

4. Spikelets borne in triads at each node . . . . **43. Hordeum**
+ Spikelets borne in pairs at each node 5

5. Raceme-rhachis tough . . . . **41. Taeniatherum**
+ Raceme-rhachis fragile . . . . **42. Crithopsis**

6. Glumes keeled to the base or sometimes rounded below 7
+ Glumes rounded on the back 8

7. Glumes acuminate, the nerves convergent at the tip . . **45. Eremopyrum**
+ Glumes obtuse to bidentate, the outer nerves separated at the tip
. . . . **46. Triticum**

8. Glumes herbaceous . . . . **39. Brachypodium**
+ Glumes coriaceous . . . . **47. Aegilops**

39. *Brachypodium*

Annual with flat, glaucous, rather stiff leaves. Weed of cultivation. Mma Md Mp Nv Di O S Sa. *Mediterranean region to C Asia* . . . . . **B. distachyum**

40. *Elymus*

1. Rhachis fragile, disarticulating between the spikelets at maturity. Coastal dunes. Mma Md Mp. *Europe, Mediterranean region and SW Asia* (see Appendix 1 note 5) . . . . **E. farctus**
+ Rhachis tough, not disarticulating at maturity 2

2. Plant tufted, without rhizomes. Damp sandy soils. Mma Md Mp. *Mediterranean region and SW Asia* . . . . **E. elongatus**
+ Plant with long creeping rhizomes, not forming tufts. Apparently not grown in Egypt although the rhizomes are sold in markets. *Europe, Mediterranean region and temp. Asia* . . . . **E. repens**

41. *Taeniatherum*

Annual with tough raceme-rhachis; awns, on drying, becoming characteristically spirally coiled. Deserts. S. *Europe and temp. Asia*
. . . . ***T. caput-medusae***

42. *Crithopsis*

Annual with fragile raceme-rhachis; awns remaining more or less straight or only slightly divergent on drying. Damp ground near wells. Mma. *E Mediterranean region eastwards to Iran* . . . . ***C. delileana***

43. *Hordeum*

1. Raceme-rhachis tough, not breaking up at maturity. Cultivated cereal (barley). Mma Nv Nn Di S. *Cult. throughout the world* . . . ***H. vulgare***
+ Raceme-rhachis fragile, readily breaking up at maturity; wild plants   2

2. Awn of fertile lemma stout, 4–14cm long. Robust annual; the ancestor of 2-rowed barley and sometimes difficult to distinguish from the self-sown crop. Desert pastures. Mma Di. *Mediterranean region to C Asia* . . . . ***H. spontaneum***
+ Awn of fertile lemma slender, not more than 3cm long 3

3. Glumes of central spikelet of the triad glabrous (*H. marinum*) 4
+ Glumes of central spikelet of the triad long-ciliate (*H. murinum*) . . . 5

4. Both glumes of the lateral spikelets subulate, or one of them slightly swollen but not winged. Sandy and alluvial soils. Mma Nv. *Mediterranean region to SW & C Asia* . . . ***H. marinum*** subsp. ***gussoneanum***
+ One glume of the lateral spikelets with a well developed wing on one side. Alluvial soils. Mma (Mp) Nv (Dl O). *Europe and the Mediterranean region to SW & C Asia* . . . . ***H. marinum*** subsp. ***marinum***

5. Leaves green; anthers of central spikelet 0.7–1mm long. Sandy fields, damp ground and canal banks. Mma Mp Nv Dl (Dg Di O S). *Mediterranean region to E Asia* . . . . . ***H. murinum*** subsp. ***leporinum***
+ Leaves glaucous; anthers of central spikelet 0.2–0.5mm long. Desert sand and as a weed of cultivation. Mma Md Mp Nv (Dl Dg) Di O S. *Mediterranean region to India* . . . . ***H. murinum*** subsp. ***glaucum***

44. *Agropyron*

Tufted perennial with laterally compressed, pectinately arranged spikelets. Sand. Mma. *C Europe, the Mediterranean region and temp. Asia* . . . . ***A. cristatum***

45. *Eremopyrum*

Glumes acute or acuminate, but not awned. Rocky soils. S. *SW and C Asia* . . . . ***E. bonaepartis***
Glumes clearly awned, the awns 2–6mm long; palea-keels produced into 2 awned teeth 0.5–1.5mm long, with a deep sinus between them. Unconfirmed in Egypt. (S). *SW and C Asia* . . . . ***E. distans***

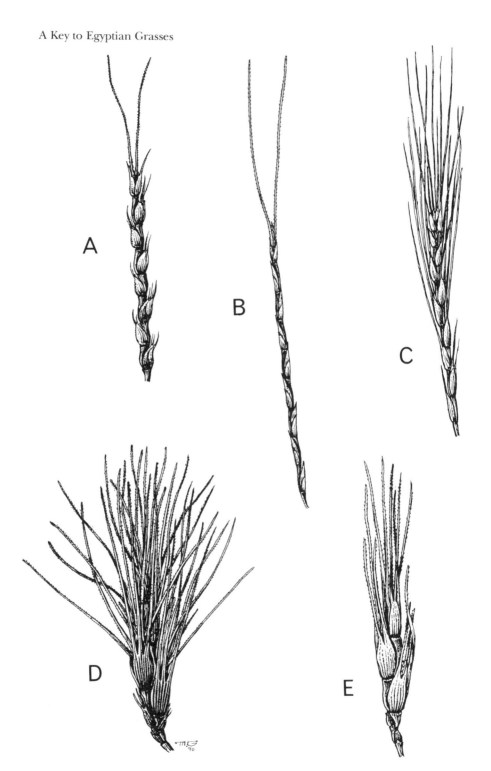

Plate 5. Racemes of: A, *Aegilops ventricosa*; B, *A. longissima*; C, *A. bicornis*; D, *A. kotschyi*; E, *A. peregrina*. Not to scale.

46. *Triticum*

1. Inflorescence-axis fragile, disarticulating at maturity. Cultivated cereal (emmer). *Cult. in many parts of the world* . . . . **T. dicoccum**
+ Inflorescence-axis tough, not disarticulating at maturity 2

2. Glumes keeled only towards the tip, rounded on the back below. Cultivated cereal (bread wheat). Mma Nv Di O. *Cult. throughout the world* . . . . **T. aestivum**
+ Glumes firmly and broadly keeled from base to tip 3

3. Raceme pyramidal, tapering towards the tip. Cultivated cereal (Egyptian cone-wheat). Mma Nv. *Cult. in Egypt and Ethiopia* **T. pyramidale**
+ Raceme parallel-sided for its whole length 4

4. Leaves glabrous. Cultivated cereal (macaroni wheat). Mma Nv Nn. *Cult. in Europe and temp. Asia* . . . . **T. durum**
+ Leaves velutinous. Cultivated cereal (rivet wheat). Nv ?Nn. *Widely cult. in temp. countries* . . . . **T. turgidum**

47. *Aegilops* (Plate 5)

1. Raceme cylindrical, 10 or more times as long as wide 2
+ Raceme ovate or lanceolate, not more than 5 times as long as wide 5

2. Raceme moniliform, contracted between the spikelets 3
+ Raceme smoothly cylindrical, not moniliform 4

3. Spikelets scabrid. Weed of barley. Mma. *Mediterranean region* **A. ventricosa**
+ Spikelets velutinous. (Mp S). *SW & C Asia* . . . . **A. crassa**

4. Terminal spikelet with 2 long, stout awns, at least one of them more than 6cm long; lateral spikelets awnless. Sandy field-margins. Mma Md Mp. *E Mediterranean region* . . . . **A. longissima**
+ Raceme awnless, or both terminal and lateral spikelets awned, the awns slender, usually not more than 6cm long. Sandy field-margins and dunes. Mma Md Mp (Nv). *Saharan and Arabian deserts* **A. bicornis**

5. Nerves of the glumes equally spaced, prominent and narrower than the spaces between, nearly parallel; awns slender. Sand. Mma Md Mp Di (O). *E Mediterranean region to the Caucasus* . . . . . . **A. kotschyi**
+ Nerves of the glumes unequally spaced, low in profile and broader than the spaces between, bowed outwards above the base; awns stout, flattened below. Sand. Mma (Mp Nv S). *Mediterranean region* **A. peregrina**

XI. ARUNDINEAE

1. Plant reed-like with leaves all cauline, or tussock-forming with harsh leaves confined to the base; panicle large, plumose 2
+ Plant not reed-like, but tufted with basal leaves; panicle not plumose 4

2. Tussock-forming plant with leaves all basal . . . . . . **50. Cortaderia**
+ Reed-like plant with leaves all cauline 3

3. Ligule a membrane; lemma hairy; rhachilla glabrous . . . **51. Arundo**
+ Ligule a line of hairs; lemma glabrous; rhachilla hairy **52. Phragmites**

4. Plant annual . . . . **48. Schismus**
+ Plant perennial from a woody rootstock . . . . **49. Centropodia**

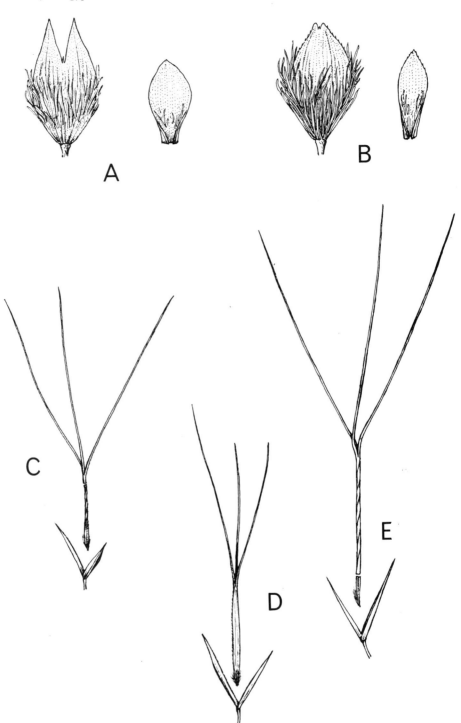

Plate 6. Lemma and palea of: A, *Schismus arabicus*; B, *S. barbatus*. Spikelets of: C, *Aristida mutabilis*; D, *A. adscensionis*; E, *A. funiculata*. Not to scale.

48. *Schismus* (Plate 6, A–B)

Lemma 2–3.5mm long, the apical lobes narrowly triangular, 0.6–1.3mm long; palea shorter than the lemma, seldom reaching beyond the middle of the lobes. Sand and gravel. Mma Md Mp Nv Dl Dg Di S Sa. *E Mediterranean region to C Asia* . . . . **S. arabicus**

Lemma 1.5–2mm long, the apical lobes broadly triangular, 0.3–0.4(0.7)mm long; palea almost as long as the lemma, often longer. Sand and gravel. Mma Md Mp Nv Dl Dg Da Di (O) S R Sa. *Mediterranean region and SW Asia; southern and southwestern Africa* . . . . **S. barbatus**

49. *Centropodia*

Robust plant up to 120cm high (seldom less than 50cm); panicle 17–35cm long; lowest lemma usually 4.8–6mm long, with an awn projecting 0.1–0.6(1.6)mm beyond the tips of the apical lobes; anthers 1.6–2.7mm long. Sand and gravel. Di. *N Africa to Arabia* . . . . **C. fragilis**

Smaller plant seldom over 30cm high; panicle 2–14cm long; lowest lemma usually 3.7–5.2mm long, with an awn projecting (0.9)1.2–2.2mm beyond the tips of the apical lobes; anthers 0.7–1.2mm long. Gravel and stable dunes. Mma Md Mp Nv Nn Dl (Dg Da) Di (O) S (R) Sa. *Trop. and northern Africa to C Asia* . . . . **C. forskalii**

50. *Cortaderia*

Large, tussock-forming perennial with harsh leaves confined to the base; plants gynodioecious. Cultivated as an ornamental (Pampas grass). *Native of S America* . . . . **C. selloana**

51. *Arundo*

Tall reed-like perennial up to 5m high with creeping woody rhizomes and large plumose panicle. Planted along water-courses, rarely occurring as a native. Mma Md Mp Nv (Nn Dl) Di O S. *Mediterranean region to SE Asia* . . . . **A. donax**

52. *Phragmites*

1. Leaves scabrid beneath (at least in the upper part), the tips attenuate, stiff; rhachilla-hairs 4–7mm long; upper glume 3–5mm long. Along water-courses. Nv. *Trop. Africa* . . . . **P. mauritianus**
+ Leaves smooth beneath, the tips filiform and flexuous; rhachilla-hairs 8–12.5mm long; upper glume 5.5–9mm long (*P. australis*) 2

2. Stems up to 4m high; panicle 15–20(30)cm long; upper glume lanceolate, sharply acute, usually apiculate. Along water-courses, in areas of cultivation and on sandy plains with seasonally high water-table. Mma Md Mp Nv Nn Dl Dg Da Di O (S) R. *Throughout the world in temp. regions* . . . . **P. australis** subsp. **australis**
+ Stems up to 6m high; panicle 30–45cm long; upper glume narrowly elliptic-oblong, obtuse to tridenticulate. Along water-courses. Mma Nv Nn Dl Di. *Shores of the Mediterranean, eastwards to Iran and southwards to Kenya and the southern fringe of the Sahara* **P. australis** subsp. **altissimus**

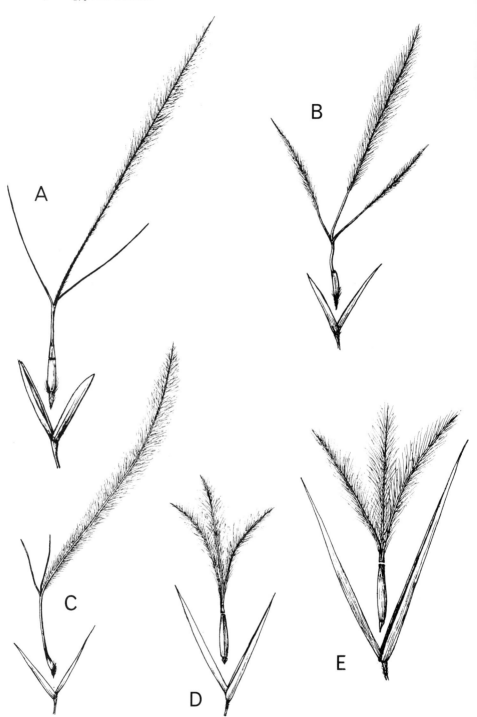

Plate 7. Spikelets of: A, *Stipagrostis ciliata*; B, *S. lanata*; C, *S. raddiana*; D, *S. vulnerans*; E, *S. scoparia*. Not to scale.

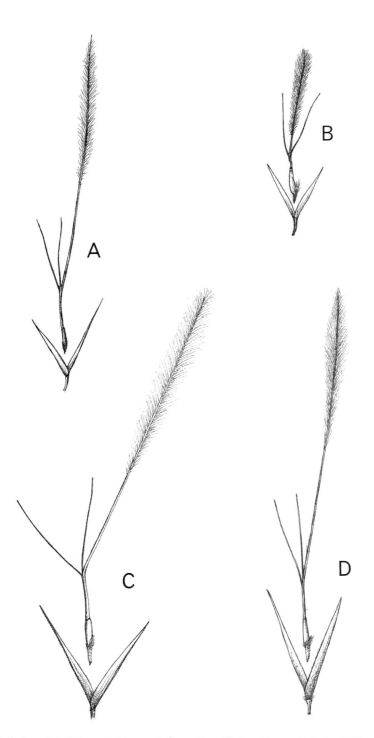

Plate 8. Spikelets of: A, *Stipagrostis plumosa*; B, *S. acutiflora*; C, *S. multinerva*; D, *S. shawii*. Not to scale.

XII. ARISTIDEAE

Central awn, and sometimes also the laterals, plumose   **53. *Stipagrostis***
All three awns glabrous . . . . **54. *Aristida***

53. *Stipagrostis* (Plates 7–9)

1. Body of lemma articulated at or just above the middle, the awns and column breaking off with the conical upper part of the lemma. Densely tufted perennial with conspicuously bearded nodes. Sandy and stony soils. Mma Md Mp Nv Dl (Dg Da) Di S R (Sa). *Macaronesia and N Africa to SW Asia* . . . . *S. ciliata*
+ Body of lemma articulated at the summit, the awns and column breaking off cleanly without a part of the lemma 2

2. All three awns plumose, at least in the upper part 3
+ Only the central awn plumose, the laterals quite glabrous 5

3. Internodes, or at least the lower, densely woolly; awns plumose only in the upper part or the central hairy almost to the base. Densely tufted perennial. Sand dunes and palm groves. Mma Md Mp Nv Dl Di (O). *N Africa and SW Asia* . . . . *S. lanata*
+ Internodes glabrous, or at most minutely pubescent 4

4. Lower glume longer than the upper. Hummock-forming suffrutescent perennial with flexuous inrolled pungent leaves. Desert sand. Mma Md Mp Nv Dl (Dg) Di O. *N Africa and SW Asia* . . . . . . *S. scoparia*
+ Lower glume shorter than the upper. Hummock-forming suffrutescent perennial with rigid inrolled pungent leaves; panicle-branches bearded in the axils; pedicels pubescent. Desert sand. (Mma) Nv Nn (Dl) Dn (Da) Di O (S). *Libya* . . . . *S. vulnerans*

5. Internodes, or at least the lower, densely woolly 6
+ Internodes glabrous, scaberulous or minutely pubescent 11

6. Central awn plumose to the base 7
+ Central awn glabrous in the lower part 8

7. Panicle lax, the spikelets spreading on long slender pedicels; central awn c.2cm long; column 2–3mm long. Densely tufted perennial. Desert sand. Di. *Saudi Arabia and Iraq* . . . . *S. drarii*
+ Panicle narrow, dense, the spikelets on short stiff pedicels; central awn 3.5–5cm long; column 8–11mm long. Densely tufted perennial. Sandy and rocky deserts. Nv (Dg Da Di O) S Sa. *N Africa and SW Asia* . . . . *S. raddiana*

8. Glumes 1- to 3-nerved 9
+ Lower glume 5- to 9-nerved; upper glume 3- to 5-nerved 10

9. Glumes 8–9mm long; lemma (including the callus) c.5mm long; central awn c.1.5cm long; column less than 1mm long. Slender perennial branched from most of the nodes. Desert sand. Nv Nn Dl Dn (Dg Da) Di (S R) Uw. *N Africa and Arabia* . . . . *S. acutiflora*
+ Glumes 15–18mm long; lemma (including the callus) 6–7mm long; central awn 2.5–6.5cm long; column up to 10mm long. Densely tufted perennial. Sandy and stony soils. Mma Md Mp Nv Nn Dl Dn (Dg) Da Di O S R Sa. *Mediterranean region and SW Asia* . . . . *S. plumosa*

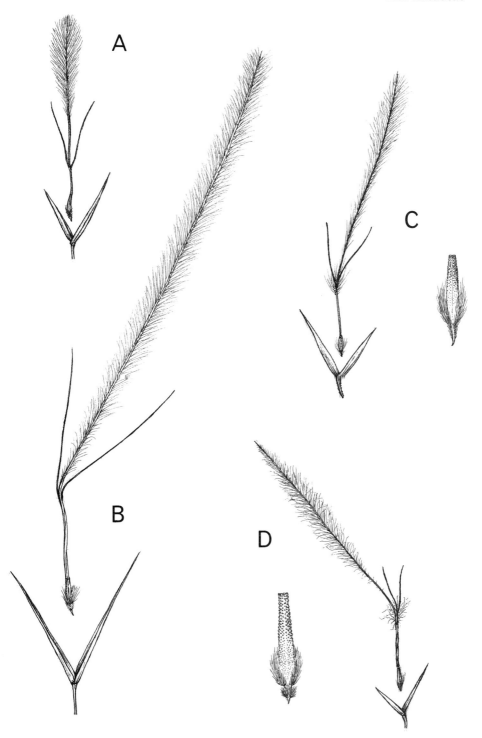

Plate 9. Spikelets of: A, *Stipagrostis obtusa*; B, *S. paradisea*; C, *S. uniplumis*; D, *S. hirtigluma*. Not to scale.

10. Loosely tufted annual. (Uw). *N Africa* . . . . *S. shawii*
+ Densely tufted perennial. (S). *Arabian Peninsula* . . . . *S. multinerva*

11. Column hairy 12
+ Column glabrous 13

12. Callus with 2 collars of hair, a lower of very short hairs just behind the
tip of the callus, and an upper of much longer hairs at the base of the
lemma-body, the callus glabrous between the collars. Annual or short-
lived perennial; panicle often lax; glumes hairy. Desert sand and rocky
mountains. Da O S R Sa. *Trop. & southern Africa to Arabia and India*
. . . . *S. hirtigluma*
+ Callus with one continuous collar of hair, the hairs increasing in
length upwards. Short-lived perennial; glumes hairy or glabrous.
Wadi-beds. Da Sa. *Trop. & southern Africa to Arabia and Pakistan*
. . . . *S. uniplumis*

13. Central awn plumose to the base, 6–7cm long. Densely tufted
perennial. The distinction between this species and *S. raddiana* needs
to be re-examined. Nv Dg. *Somalia and SW Asia* . . . . . *S. paradisea*
+ Central awn glabrous in the lower half, 2–3cm long. Densely tufted
perennial with short stiff curved leaves confined to a short basal
cushion. Sands and gravels. Mma Nv (Dg) Di (S). *N Africa to SW Asia;
South Africa* . . . . *S. obtusa*

54. *Aristida* (Plate 6, C–E)

1. Lemma or column not articulated at the summit. Annual or short-lived
perennial. An ubiquitous weed. Nv (Nn Dl) Dg (Da) Di O S R Sa.
Throughout the tropics and subtropics . . . . *A. adscensionis*
+ Lemma or column articulated at the summit 2

2. Articulation at the top of the column, just below the awns. Annual.
Sand. Da (S) R Sa (Uw). *Trop. Africa to Arabia and India* . . *A. mutabilis*
+ Articulation at the base of the column. Annual. Sand. (Mma Da) R Sa
(Uw). *Trop. Africa to Arabia and India* . . . . *A. funiculata*

XIII. PAPPOPHOREAE

Lemma 9-awned . . . .**55. Enneapogon**
Lemma 5-awned, the awns alternating with 6 hyaline lobes **56. Schmidtia**

55. *Enneapogon* (Plate 10, A–D)

1. Awns of fertile lemma scaberulous throughout. Stony ground. Sa.
N Africa and Somalia; southern Africa . . . . *E. scaber*
+ Awns of fertile lemma ciliate for most of their length, scaberulous only
towards the tip .2

2. Fertile lemma with 3 dense patches of hair on the back, one along the
midnerve and one along each margin. Annual; basal sheaths
remaining intact. Wadi-beds. Sa. *Ethiopia, Somalia and Arabia*
. . . . *E. lophotrichus*
+ Fertile lemma with hairs on the back evenly distributed; tufted
perennials, rarely annual 3

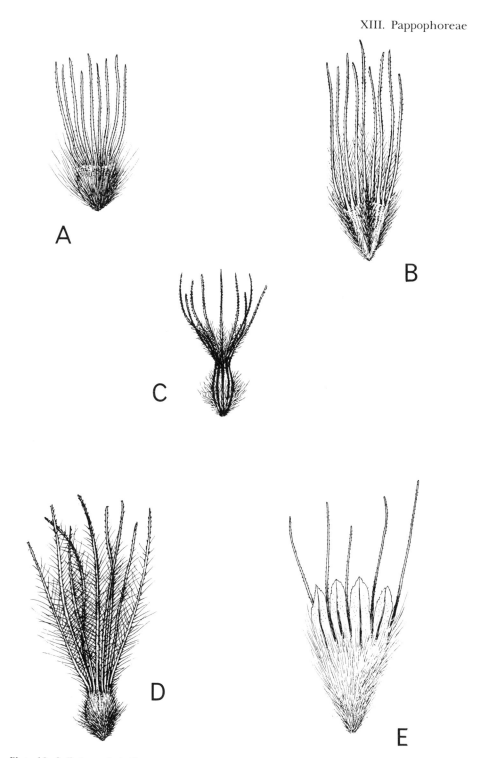

Plate 10. Spikelets of: A, *Enneapogon scaber*; B, *E. lophotrichus*; C, *E. desvauxii*; D, *E. persicus*; E, *Schmidtia pappophoroides*. Not to scale.

3. Basal sheaths persistent, forming a bulbous base to the stem, eventually disintegrating into a cushion of fibres; spikelets 3.5–6mm long; third lemma vestigial, 0.3–0.8mm long. Sandy soils and rock fissures. (Da) S (R) Sa. *Africa to India and China; C & S America* . . . . ***E. desvauxii***
+ Basal sheaths neither forming a bulbous base to the stem nor disintegrating into a cushion of fibres; spikelets 6–13mm long; third lemma well developed, 2.5–9(10.5)mm long, often accompanied by a vestigial fourth. Rocky ground. Nv Sa. *Trop. Africa to S & SW Asia and India* . . . . ***E. persicus***

56. *Schmidtia* (Plate 10, E)

Shortly rhizomatous perennial, often with long surface stolons; leaves pubescent. Sandy ground. O. *Trop. and southern Africa, Socotra and Pakistan* . . . . ***S. pappophoroides***

XIV. ERAGROSTIDEAE

1. Spikelets 2- to several-flowered 2
+ Spikelets 1-flowered 15

2. Lemmas 9- to 11-nerved . . . . **57. *Aeluropus***
+ Lemmas 3-nerved (if clavate hairy on the back below see 17. *Desmazeria*), sometimes 1 or more subsidiary nerves present on either side of the keel in 66. *Eleusine* 3

3. Tip of lemma emarginate to 2- to 3-lobed, or the flanks hairy between the lateral nerves and the margin, or florets conspicuously bearded from the callus 4
+ Tip of lemma entire, the nerves and flanks glabrous; florets not bearded from the callus 11

4. Grain strongly flattened, concavo-convex, with a free pericarp . . . . **65. *Coelachyrum***
+ Grain seldom flattened and then not with a free pericarp 5

5. Inflorescence a panicle . . . . **58. *Triraphis***
+ Inflorescence comprising 2 or more racemes 6

6. Racemes persistent 7
+ Racemes deciduous 10

7. Lower glume much longer than the lowest lemma . . . **61. *Trichoneura***
+ Lower glume not exceeding the lowest lemma 8

8. Spikelets disarticulating above the glumes but not between the florets . . . . see XV. Cynodonteae
+ Spikelets disarticulating between the florets, or the rhachilla persistent 9

9. Florets bearded from the callus . . . . **60. *Halopyrum***
+ Florets not bearded from the callus . . . . **59. *Leptochloa***

10. Glumes as long as the spikelet, enclosing the florets; racemes arranged along a central axis . . . . **62. *Dinebra***
+ Glumes not or scarcely exceeding the adjacent lemmas; racemes digitate . . . . **63. *Ochthochloa***

57. *Aeluropus*

Inflorescence a globose, elliptic or oblong head of closely crowded
spikelets; lemmas hairy. Low-growing sward-forming perennial with rigid,
subdistichous leaves. Damp and arid places, both fresh and saline. Mma
(Md) Mp Nv Dl Dg (Da) Di O S R Sa. *Mediterranean region to India*
. . . . *A. lagopoides*
Inflorescence elongate, comprising several widely spaced racemes;
lemmas glabrous. Sward-forming perennial with rigid subdistichous
leaves. Sandy and salty soils. (Mp) S R. *Mediterranean region to E Asia*
. . . . *A. littoralis*

58. *Triraphis*

Annual with ovoid spike-like panicle; lemmas villous on the lateral
nerves, 3-lobed, 3-awned, the central lobe 2-toothed. Sandy and stony
ground. (Da R) Sa. *Trop. Africa and Arabia* . . . . *T. pumilio*

59. *Leptochloa*

Spikelets 6–15mm long, subterete with rounded lemma, loosely
arranged in the indistinctly secund racemes. Rhizomatous or more or less
tufted perennial up to 150cm high. Watersides, saltmarshes and as a weed
of rice. Mma Md Mp Nv Nn Di O R Sa. *Trop. and subtrop. Old World* **L.** *fusca*
Spikelets 2–3mm long, laterally compressed with keeled lemma,
overlapping in the clearly secund racemes. Annual up to 110cm high.
Alluvium. Nv. *Trop. Africa and trop. Asia* . . . . *L. panicea*

60. *Halopyrum*

Tough, tussocky perennial with woody stems up to 2m high; callus and
rhachilla-tip bearded with white hairs half as long as the lemma. Coastal
dunes, growing up through successive accretions of sand. (R) Sa. *Shores of
the Indian Ocean from Moçambique to Sri Lanka* . . . . . . **H.** *mucronatum*

61. *Trichoneura*

Slender annual with fairly compact panicles of usually ascending racemes; glumes exceeding lowest lemma. Sand. (Da R) Sa. *Trop. Africa and Arabia* . . . . **T. mollis**

62. *Dinebra*

Annual with reflexed linear or wedge-shaped racemes scattered along a central axis. Damp soils and as a weed of maize, cotton and sugar-cane. Mma Md Mp Nv O Sa. *Trop. Africa, through Arabia to India* . . . **D. retroflexa**

63. *Ochthochloa*

Sprawling stoloniferous perennial with digitate deciduous racemes. Unconfirmed in Egypt. (Da R Sa). *NE Africa to NW India* . . . **O. compressa**

64. *Eragrostis* (Plate 11)

1. Creeping perennial; panicle narrow, contracted. Probably introduced. (Mp). *Southern Africa and trop. Asia; intr. in Palestine* . . . **E. sarmentosa**
+ Annual or tufted perennial 2

2. Florets persistent and the grain retained on the mature panicle. Cultivated cereal or fodder grass (tef) from Ethiopia, tested as a forage crop in Egypt and occasionally found as a casual. Nv O. *Originally from Ethiopia* . . . . **E. tef**
+ Florets variously breaking up at maturity 3

3. Spikelets shedding their florets from the tip downwards, the rhachilla-segments falling with the florets 4
+ Spikelets shedding their florets from below upwards, the rhachilla tough, remaining on the pedicel, or eventually breaking up 8

4. Palea ciliate on the keels, the long hairs clearly visible beyond the margins of the lemma 5
+ Palea glabrous or scabrid on the keels 7

5. Panicle contracted, spike-like or lobed and interrupted, the spikelets densely crowded; anthers 2. Sandy soils. (Da R) Sa. *Throughout the tropics* . . . . **E. ciliaris**
+ Panicle loose and open 6

6. Anthers 2; glumes lanceolate, unequal; lemmas oblong, truncate and mucronate. Rocky places. Sa. *E trop. Africa and Arabia* . . . . **E. lepida**
+ Anthers 3; glumes ovate, subequal; lemmas ovate, rounded at the tip. Likely to occur, but all material seen thus far is referable to *E. lepida*. *Throughout the tropics* . . . . **E. tenella**

7. Ligule a membrane; lemmas 0.7–1mm long. Annual or short-lived perennial with narrow panicle of small spikelets. Sandy and heavy soils. Nv Nn (S). *Trop. Africa to SE Asia* . . . . **E. japonica**
+ Ligule a line of hairs; lemmas 1.1–1.5mm long. Panicle very diffuse with stiffly spreading branches. Damp sandy soils. Sa. *Trop. and South Africa to India* . . . . **E. aspera**

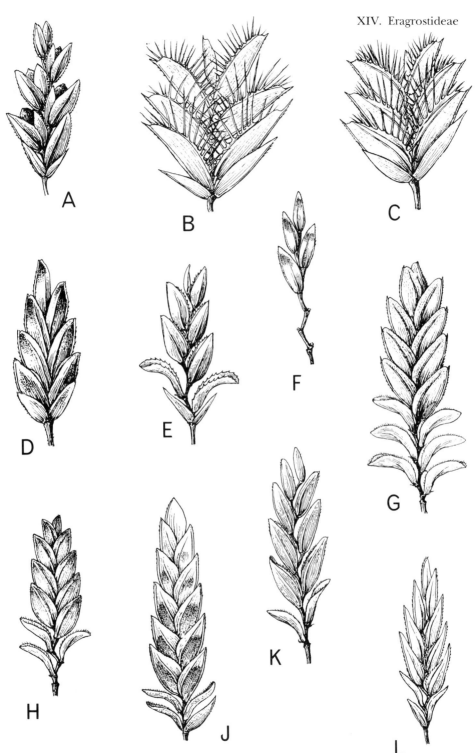

Plate 11. Spikelets of: A, *Eragrostis tef*; B, *E. ciliaris*; C, *E. lepida*; D, *E. japonica*; E, *E. aspera*; F, *E. pilosa*; G, *E. cilianensis*; H, *E. minor*; J, *E. tremula*; K, *E. barrelieri*; L, *E. aegyptiaca*. Not to scale.

8. Palea falling at about the same time as the lemma, leaving the bare persistent, zig-zag rhachilla. Panicle-branches bearded in the axils. Weed of cultivation. Mma Nv Nn Dl (Da) Di O (Sa). *Trop. and warm temp. regions around the world* . . . . *E. pilosa*
+ Palea persisting on the rhachilla long after the lemma has fallen, sometimes the rhachilla disarticulating above the glumes at an early stage with most of the florets still attached, or the upper part becoming fragile once the lower lemmas have begun to fall 9

9. Plant perennial, up to 1m high. Cultivated in Giza in 1976 but it is not known whether it is still there. *Native of trop. Africa* . . . . *E. tenuifolia*
+ Plants annual 10

10. Grain subrotund, 0.4–0.6mm in diameter; leaf-margins with or without crateriform glands. A variable species with a generally ovate, rather dense panicle, intergrading with *E. minor* and best distinguished by the shape of the grain. Damp sand and areas of cultivation. Mma Md Mp Nv Nn Dl (Dg) Da Di O (S) R Sa. *Trop. and warm temp. Old World* . . . . *E. cilianensis*
+ Grain broadly oblong to elliptic-oblong, 0.5–1mm long, if subrotund and 0.4–0.6mm in diameter then spikelets trembling on long fine pedicels 11

11. Leaves with crateriform glands along the margins. Rather similar to *E. cilianensis* but with more open panicle, narrower oblong rather than ovate spikelets, shorter lemmas and different shape of grain. Damp soils. (Nv O) S. *Subtrop. and warm temp. Old World* . . . . . *E. minor*
+ Leaves without crateriform glands along the margins 12

12. Spikelets trembling on long fine pedicels in a very loose open panicle. An occasional introduction. (Nv*). *Trop. Africa and India* . *E. tremula*
+ Spikelets on short stiff pedicels in a more or less contracted, sometimes open, panicle 13

13. Panicle-branches solitary or paired; lemmas 1.7–2.3mm long, obtuse. Spikelets usually tinged with purple, in a more or less open panicle. Watersides and cultivated fields. Mma Nv Nn Dl Di O S Sa. *Mediterranean region and trop. Africa to SW Asia* . . . . . *E. barrelieri*
+ Panicle-branches, or at least the lower, clustered or whorled; lemmas 1.4–1.6mm long, subacute. Spikelets usually pallid in a rather dense, but sometimes spreading, panicle. Along water-courses. Mma Nv Nn (Dl Da) Di Sa (Uw). *Endemic to Egypt* . . . . *E. aegyptiaca*

65. *Coelachyrum*

Tufed annual with subdigitate racemes; lemmas inconspicuously hairy. Sandy and rocky soils. (Da R) Sa. *N & NE Africa and Arabia* . . *C. brevifolium*

66. *Eleusine*

1. Perennial; midnerve of lemma simple; leaves with tufts of hair scattered along the margins. Densely tufted and sometimes mat-forming from a tough branching rhizome. Introduced and grown for basket making, occasionally found as an escape. Nv*. *Ethiopia, Somalia and Arabia* . . . . *E. floccifolia*
+ Annuals; midnerve of lemma with 2–3 subsidiary nerves close to each side of it forming a thickened keel 2

2. Racemes 9–15mm wide; spikelets ovate, non-shattering, very closely overlapping; grain plump, almost globose, usually brown, often exposed between the gaping lemma and palea when ripe. Cultivated for forage and occasionally escaping. (Mma) Nv. *Cult. in the Old World tropics and subtropics* . . . . ***E. coracana***
+ Racemes 3–7mm wide; spikelets elliptic, disarticulating between the florets; grain oblong to broadly oblong, blackish, never exposed when ripe; wild grass (*E. indica*) 3

3. Slender plant; ligule sparsely and minutely ciliolate; racemes 3–5.5mm wide; lower glume 1-nerved, 1.1–2.3mm long; upper glume 1.8–2.9mm long; lemmas 2.4–3.6(4)mm long; grain obliquely striate with very fine close perpendicular lines running between the striae; diploid. Weed of cultivation. (Mma) Mp Nv (Di S). *Pantropical*   ***E. indica*** subsp. ***indica***
+ Robust plant; ligule with a definite ciliolate fringe; racemes 4–7mm wide; lower glume often 2- to 3-nerved, 2–3.2(3.9)mm long; upper glume 3–4.7mm long; lemmas 3.7–4.9mm long; grain obliquely ridged, uniformly granular; tetraploid. Weed of cultivation. Mma Mp Nv Di O R. *Uplands of eastern and southern Africa*   . . . ***E. indica*** subsp. ***africana***

67. *Acrachne*

Annual with subdigitate or scattered racemes; lateral nerves of lemma shortly excurrent. An introduced weed of maize. Nv*. *Old World tropics*
. . . . ***A. racemosa***

68. *Dactyloctenium*

1. Stoloniferous perennial. Racemes 0.8–2cm long, slightly falcate. Sand. (Nn Da R) Sa. *NE Africa through Arabia to NW India* . . .   ***D. scindicum***
+ Annuals, sometimes stoloniferous 2

2. Inflorescence open, the racemes 1.2–6.5(7.5)cm long, linear to narrowly oblong, ascending or radiating; lemmas acute, cuspidate or mucronate; grain transversely rugose. Weed of cultivation. Mma Md Mp Nv Nn Dl (Dg) Di O S. *Trop. and warm temp. Old World*   ***D. aegyptium***
+ Inflorescence compact, the racemes 0.8–1.8(2.6)cm long, oblong to broadly oblong, clustered in a dense head; lemmas conspicuously acuminate-mucronate; grain granular or granular-striate. Saline sandy soils near the coast. Sa. *E Africa through Arabia to NW India*
. . . . ***D. aristatum***

69. *Desmostachya*

Harsh tussocky rhizomatous perennial up to 120cm high; inflorescence an erect linear panicle comprising short racemes. By springs, along water-courses, in wet alluvium and in areas of cultivation. Mp Nv Nn (Dl Dg) Da Di O (S). *Old World tropics* . . . . ***D. bipinnata***

70. *Sporobolus*

1. Panicle-branches, or at least the lower, in whorls. Tussocky perennial, often stoloniferous, with stiff pungent leaves and ovate to pyramidal panicle. Desert sand, but unconfirmed in Egypt. *Trop. Africa through Arabia to India* . . . . ***S. ioclados***
+ Panicle-branches not whorled, not even the lowermost 2

2. Leaves pungent, tightly inrolled; panicle spike-like 3
+ Leaves not pungent 4

3. Panicle smoothly cylindrical; leaves stiff and pungent but not conspicuously distichous. Harsh, spiky, mat-forming perennial. Seasonally inundated, saline desert sand. Mma (Mp) Nv Nn Dl (Dg) Da (Di) O S (R) Sa. *Africa to India* . . . . *S. spicatus*
+ Panicle untidily ovate, embraced below by the uppermost sheath; leaves stiff and pungent and conspicuously distichous. Sandy seashores and saltmarshes. Mma Mp (Di). *Mediterranean region* *S. pungens*

4. Upper glume ½ as long as the spikelet. Robust perennial up to 1.2m high. Introduced to Fuka in the 1950s, presumably on trial as a fodder grass, but probably no longer there. Mma*. *Trop. and southern Africa and Arabia* . . . . *S. natalensis*
+ Upper glume ⅔–¾ as long as the spikelet. Robust perennial up to 2m high. Introduced to Ras el Hekma in the 1950s, presumably on trial as a fodder grass, but probably no longer there. Mma*. *USA and Mexico* . . . . *S. wrightii*

71. *Crypsis*

1. Inflorescence at least 5 times as long as wide, scarcely embraced by the uppermost sheath which itself is hardly inflated. Cultivated ground and roadsides. (Mma Mp) Nv (Nn) Di. *Mediterranean region to C Asia* . . . . *C. alopecuroides*
+ Inflorescence not more than twice as long as wide, tightly embraced below by a pair of conspicuously inflated sheaths bearing reduced or rudimentary blade 2

2. Blade of uppermost leaf continuous with its sheath; inflorescence capitate, wider than or as wide as long; palea 1-nerved. Cultivated ground. (Mma Mp) Nv (Nn). *Mediterranean region to E Asia* . *C. aculeata*
+ Blade of uppermost leaf clearly demarcated from its sheath; inflorescence ovoid, longer than wide; palea 2-nerved 3

3. Collar and sheath-margins glabrous; glumes unequal, shorter than the lemma; anthers c.0.9mm long. (Mma Mp) Nv (Nn Da Di O R). *Europe, Mediterranean region and temp. Asia* . . . . *C. schoenoides*
+ Collar and sheath-margins hairy; glumes subequal, as long as the lemma; anthers 0.6–0.7mm long. Wet soils. Nv Nn Dl Dg. *Trop. Africa to India* . . . . *C. vaginiflora*

XV. CYNODONTEAE

1. Spikelets containing 2–5 fertile florets . . . . **72. *Tetrapogon***
+ Spikelets containing 1 fertile floret 2

2. Racemes persistent, the spikelets breaking up at maturity 3
+ Racemes deciduous or spikelets falling entire 6

3. Spikelets 1-flowered, awnless . . . . **75. *Cynodon***
+ Spikelets 2- to several-flowered, or the lemma sinuously awned . . . 4

4. Grain and lemma dorsally compressed . . . . . . **74. *Enteropogon***
+ Grain trigonous to subterete; lemma laterally compressed 5

5. Lemma produced into a long sinuous awn . . . . . **76. Schoenefeldia**
+ Lemma with a short straight awn . . . . **73. Chloris**

6. Spikelets 2-flowered in a cuneate raceme . . . . . **77. Melanocenchris**
+ Spikelets 1-flowered 7

7. Lower glume very small or suppressed; upper glume and lemma about
   equal, with raised nerves bearing hooked bristles . . . . **78. Tragus**
+ Lower glume well developed, often modified into a long flat recurved
   tail; upper glume usually smaller, enfolding the lemma **79. Leptothrium**

72. *Tetrapogon*

Glumes inconspicuous, 2–4mm long. Densely tufted perennial with conspicuously keeled and tightly flabellate basal sheaths; uppermost sheath only slightly inflated. Rocky hillsides. (Mp Dg Da) Di S Sa. *N and NE Africa through Arabia to India* . . . . *T. villosus*
Glumes conspicuous, 5–12mm long. Annual or short-lived perennial, the basal sheaths keeled and loosely flabellate; uppermost sheath very inflated. Rocky ground. (Sa). *Macaronesia, trop. Africa and Arabia* . . . . **T. cenchriformis**

73. *Chloris*

1. Leaves obtuse. Stoloniferous annual with keeled sheaths. Garden weed. Nv*. *Trop. Africa and Arabia; S America* . . . . . . **C. pycnothrix**
+ Leaves tapering 2

2. Fertile lemma obliquely obovate in profile, the keel slightly gibbous, with a crown of long spreading hairs at the tip. Annual. Sandy ground beside water and in areas of cultivation. Nv Nn (O) Sa. *Throughout the tropics* . . . . **C. virgata**
+ Fertile lemma lanceolate in profile, not gibbous, without a crown of spreading hairs at the tip though ciliate on the margins. Stoloniferous perennial. Garden escape now established as a field weed. Mma* Nv* Nn* O*. *Trop. and southern Africa* . . . . **C. gayana**

74. *Enteropogon*

Annual with digitate racemes; lemma dorsally compressed, otherwise the plant resembles a species of *Chloris*. Sand. (Da R) Sa. *Macaronesia, trop. Africa and Arabia* . . . . **E. prieurii**

75. *Cynodon*

Leaves filiform, 1–4cm long, up to 1mm wide; racemes usually 2, 0.7–1.5cm long. A delicate species of low stature (up to 15cm high) cultivated as a lawn grass. *Trop. and southern Africa* . . . . **C. transvaalensis**
Leaves flat or inrolled, up to 12cm long and 4mm wide; racemes usually 4–6, 1.5–6(8)cm long. Sward-forming perennial. Sandy and rocky soils, along water-courses and by lakes, and as a weed of cultivation. Mma Md Mp Nv Nn Dl (Dg Da) Di O S R Sa. *Trop. and warm temp. regions* **C. dactylon**

76. *Schoenefeldia*

Annual; recognised at once by the beautifully braided awns. Sand. (Nn). *Trop. Africa through Arabia to Pakistan and India* . . . . **S. gracilis**

77. *Melanocenchris*

Annual with 3–5 cuneate deciduous racemes, these 10mm or more long, including the awns, and 6–14mm apart. Sand. (Da) Sa. *NE Africa through Arabia to NW India* . . . . *M. abyssinica*

78. *Tragus*

Spikelets in a racemelet of 2–4 fertile and 1–2 sterile, separated by distinct internodes, the rhachis prolonged; upper glume 7-nerved. Dry sandy or stony ground. (Da R) Sa. *Europe, temp. Asia and northern trop. Africa* . . . . *T. racemosus*
Spikelets in pairs, those of a pair separated by a distinct internode, the upper shorter than the lower, the rhachis not prolonged; upper glume 5-nerved. Wadi-beds. (Da) R Sa. *Africa, SW Asia, China and America* . . . . *T. berteronianus*

79. *Leptothrium*

Short-lived perennial; glumes tuberculate-spinulose, the lower usually modified into a long flat recurved tail. Sand. (Da R) Sa. *Trop. Africa and SW Asia* . . . . *L. senegalense*

XVI. PANICEAE

1. Spikelets not subtended by bristles or spines 2
+ Spikelets, or some of them, subtended by bristles or spines 10

2. Inflorescence an open or spike-like panicle, sometimes condensed about the primary branches 3
+ Inflorescence consisting of 1-sided racemes, these either digitate or scattered along a central axis; rarely the spikelets long-pedicelled and distant (*Brachiaria deflexa*) 5

3. Upper lemma coriaceous to crustaceous; lower lemma entire at the tip, awnless; lower glume more or less ovate, distinct . . . . **80. Panicum**
+ Upper lemma cartilaginous; lower lemma often 2-lobed at the tip or awned; lower glume oblong, very small or almost suppressed 4

4. Upper floret dorsally compressed, the stigmas emerging terminally . . . . **87. Tricholaena**
+ Upper floret laterally compressed, the stigmas emerging laterally . . . . **88. Melinis**

5. Racemes very short, of 1–8 spikelets, more or less embedded in the thickened axis . . . . **86. Stenotaphrum**
+ Racemes free, appressed or divergent, but not sunk in a thickened axis 6

6. Upper lemma chartaceous to cartilaginous, with thin flat margins covering most of the palea and often overlapping . . . . **89. Digitaria**
+ Upper lemma coriaceous to crustaceous, with narrow inrolled margins clasping only the edge of the palea 7

7. Lower glume absent . . . . **83. Paspalum**
+ Lower glume present 8

8. Racemes mostly 4-rowed, the spikelets in clusters of 2 or more; spikelets gibbously plano-convex, cuspidate to awned; upper lemma acute, awnless; upper palea acute, with reflexed tip . . . **81. *Echinochloa***
+ Racemes mostly 1- to 2-rowed, the spikelets single or paired 9

9. Upper palea acute, its tip often slightly reflexed; racemes short, compact, appressed to the long, slightly hollowed common axis; spikelets borne singly on the racemes, the lower glume turned away from the axis . . . . **85. *Paspalidium***
+ Upper palea obtuse, its tip not reflexed; racemes more or less spreading; spikelets borne singly or in pairs . . . . . **82. *Brachiaria***

10. Bristles persisting on the axis after the spikelets have fallen **84. *Setaria***
+ Bristles or spines falling with the spikelets 11

11. Bristles free throughout, more or less filiform . . . . . **90. *Pennisetum***
+ Bristles or spines flattened and united below, commonly forming a disc or cup . . . . **91. *Cenchrus***

80. *Panicum*

1. Upper lemma faintly to strongly rugose. Densely tufted perennial up to 3m high. Cultivated for fodder (Guinea grass). Nv*. *Trop. and southern Africa* . . . . ***P. maximum***
+ Upper lemma quite smooth 2

2. Spikelets 4–4.5mm long 3
+ Spikelets up to 3.5mm long 4

3. Sheaths hairy; spikelets persistent; lower glume acuminate, ½ to ¾ the length of the spikelet. Robust hispid annual. Rarely cultivated as a cereal, but readily escaping. (Mma*) Nv*. *Cult. in warm temp. regions; originally from India* . . . . ***P. miliaceum***
+ Sheaths glabrous; spikelets deciduous; lower glume about as long as the spikelet. Glabrous twiggy perennial froming rounded bushy clumps 1m or more high. Deserts. Mma Md Mp Nv Nn Dl Dg Da Di (O) S R Sa. *N & NE Africa and SW Asia* . . . . ***P. turgidum***

4. Spikelets narrowly lanceolate to oblong-lanceolate, widest below the middle, acute. Tough rhizomatous perennial with stems up to 1m high arising from a knotty base. Canal-banks, ditches and rice-fields, and in coastal sand. Mma Md Mp Nv Nn (Dg) Di O. *Tropics and subtropics* . . . . ***P. repens***
+ Spikelets oblong, ovate-oblong or elliptic, widest at about the middle 5

5. Lower glume ¼ to ½ the length of the spikelet; margins of the glumes herbaceous. Perennial with knotty or slightly swollen base. Canal-banks and gardens. Mma Mp Nv (Nn) Di (O). *Trop. Africa and Arabia* . . . . ***P. coloratum***
+ Lower glume ½ to ¾ the length of the spikelet; margins of the glumes broadly membranous. Perennial arising from a creeping woody rootstock. Introduced, presumably for fodder, but possibly native in Sinai. Mma* Nv* R. *Trop. Africa to India* . . . . ***P. antidotale***

81. *Echinochloa*

1. Ligule represented by a line of hairs, at least in the lower leaves; plants perennial 2

+ Ligule quite absent; plants annual 3

2. Spikelets awnless or rarely with a subulate point up to 3mm long,
plump, 2.5–4mm long; stems robust, erect, reed-like. Unconfirmed in
Egypt. *Trop. and southern Africa and Arabia* . . . . . . *E. pyramidalis*
+ Spikelets awned, the awns up to 20(50)mm long, 3.5–6mm long;
rhizomatous plant with spongy stems decumbent and rooting at the
nodes. Invasive weed of canals and ditches. Mma (Md) Mp Nv (Di O).
Trop. Africa; India to Indo-China . . . . *E. stagnina*

3. Spikelets acuminate to awned, 3–4mm long, hispid; racemes untidily
2–several-rowed, the longest 2–10cm long and usually with secondary
branchlets at the base. Weed of irrigation channels and rice-fields.
Mma Md Mp Nv Dl O. *Subtropics* . . . . *E. crusgalli*
+ Spikelets acute, 1.5–3mm long, pubescent; racemes neatly 4-rowed,
openly spaced, rarely more than 3cm long, simple. Canal-banks,
gardens and cultivated fields. Mma Md Mp Nv Nn (Dl Dg Da Di) O S (R
Sa). *Tropics and subtropics* . . . . *E. colona*

82. *Brachiaria*

1. Plants perennial. Racemes with flat, more or less ribbon-like narrowly
winged rhachis. Canal-banks and ditches. Mma Mp Nv (Di). *Tropics*
. . . . *B. mutica*
+ Plants annual 2

2. Upper lemma smooth, shining, obtuse; spikelets neatly imbricate;
lower glume up to ⅓ the length of the spikelet. Weed of cultivation.
(Mp) Nv Nn O S. *Mediterranean region to southern Africa and India*
. . . . *B. eruciformis*
+ Upper lemma granulose to rugose, subacute to mucronulate 3

3. Glumes separated by a distinct internode; upper lemma coarsely
rugose; racemes slender, distant, secund. Wadis. (Da) Sa. *Trop. Africa
and Arabia* . . . . *B. leersioides*
+ Glumes adjacent, or if slightly separated then the upper lemma not
coarsely rugose 4

4. Racemes compound, bearing the spikelets in dense fascicles or on
short secondary branchlets. Sandy wadis and seashores. (Da R Sa).
Trop. and southern Africa to Arabia and India . . . . *B. deflexa*
+ Racemes simple, the spikelets borne in pairs 5

5. Spikelets 1.5–2.2mm long; lower glume up to ¼ the length of the
spikelet, usually truncate. Weed of cultivation. (Mma) Mp Nv Da. *Trop.
Asia, Arabia and trop. Africa* . . . . *B. reptans*
+ Spikelets 2.5–3.5mm long; lower glume up to ½ the length of the
spikelet. Moist sandy soils. (Da R) Sa. *Trop. and southern Africa and trop.
Asia* . . . . *B. ramosa*

83. *Paspalum*

1. Spikelets with a ciliate fringe from the margins of the upper glume.
Robust perennial up to 1.8m high. Introduced garden weed. Nv*.
Native of S America . . . . *P. dilatatum*
+ Spikelets glabrous or minutely pubescent, without a ciliate fringe . . 2

2. Leaves linear; racemes paired, rarely a third present below. Creeping stoloniferous perennial. Ditches and canal-banks, rarely in cultivated fields. (Mma) Md (Mp) Nv Nn Di O (S). *Tropics and subtropics* . . . . **P. distichum**
+ Leaves broadly lanceolate, cordate at the base; racemes numerous, crowded in a narrowly oblong head. Annual. Introduced and at one time grown as a forage crop but apparently no longer in cultivation. Nv*. *Native of S America* . . . . **P. racemosum**

84. *Setaria*

1. Panicle clearly branched; leaves pleated fanwise, especially at the base of the younger ones. Perennial forming clumps up to 3m high. Garden ornamental. *Trop. and southern Africa; trop. America* . . . **S. megaphylla**
+ Panicle spike-like, sometimes more or less lobed; leaves not at all pleated 2

2. Spikelets not deciduous, the upper lemma disarticulating at maturity above the persistent lower lemma. Rarely cultivated as a cereal (Italian foxtail millet). (Nv*). *Cult. in the warm temp. Old World* . . . . **S. italica**
+ Spikelets deciduous as a whole 3

3. Bristles retrorsely barbed, clinging tenaciously to clothing or fur; panicle often untidily lobed. Annual. Weed of cultivation. Mma Md Mp Nv Nn (Dg) Da Di O S R Sa. *Trop. and warm temp. regions* . . . . **S. verticillata**
+ Bristles antrorsely barbed 4

4. Panicle untidily lobed; occasionally 1 or 2 bristles retrorsely barbed. The hybrid between *S. viridis* and *S. verticillata*. Cultivated ground and waste places. Mma Mp Nv (Dg) Di O S R Sa. *Occasional where the parents grow together* . . . . **S. x verticilliformis**
+ Panicle smoothly cylindrical 5

5. Upper glume as long as the upper lemma, the latter finely rugose. Weed of cultivation. Mp Nv Nn Di O (S). *Temp. Old World* . . **S. viridis**
+ Upper glume shorter than the upper lemma, the latter strongly rugose to corrugate, rarely almost smooth. Canal-banks and areas of cultivation. Mma (Mp) Nv Nn (Di) O (S Sa). *Trop. and warm temp. Old World* . . . . **S. pumila**

85. *Paspalidium*

Spikelets 1.6–2.6mm long; leaves setaceously acuminate. Plant creeping with spongy stolons; raceme-rhachis narrowly winged. In ditches, along canals and in swamps. Mma (Md) Mp Nv Di O. *Old World tropics* . . . . **P. geminatum**
Spikelets 3–4mm long; leaves obtuse to bluntly acute. Plant creeping or floating with spongy rhizomes; raceme-rhachis broadly winged. Ditches. (Mma Mp) Nv. *Trop. and southern Africa; Algeria* . . . . . . **P. obtusifolium**

86. *Stenotaphrum*

Stoloniferous perennial with broadly linear, blunt-tipped leaves, laterally compressed sheaths and numerous short racemes more or less embedded in the spongy inflorescence-axis. Cultivated as a lawn grass (St. Augustine grass) and may be naturalised on the coast. *Atlantic coasts of Africa and America* . . . . **S. secundatum**

87. *Tricholaena*

Tufted perennial with conspicuously white-hairy spikelets and woody stems arising from a knotty base. Sandy and stony soils. Dg Da Di S R Sa. *Macaronesia to India* . . . . **T. teneriffae**

88. *Melinis*

Spikelets 1.5–2(2.4)mm long; upper glume straight on the back. Perennial up to 1m high. Grown in the garden of the Agricultural Museum, Dokki. Nv*. *Native of trop. Africa, widely grown as a fodder grass (Molasses grass)* . . . . **M. minutiflora**
Spikelets 2.5–8.5mm long; upper glume gibbous. Annual or short-lived perennial with conspicuously hairy spikelets often tinged with pink or purple. An escape from cultivation. Nv*. *Africa* . . . . . . **M. repens**

89. *Digitaria*

1. Plants perennial. Tufted plant with hairy cataphylls at the base of the stem and pubescent to villous spikelets. Rocky ground. Sa. *N & NE Africa to Arabia and Pakistan* . . . . **D. nodosa**
+ Plants annual, usually straggling 2

2. Spikelets ternate. Nn. *Trop. Asia and trop. America* . . . **D. violascens**
+ Spikelets binate 3

3. Nerves of the lower lemma scabrid, otherwise glabrous. Weed of cultivation. Mma (Mp) Nv Nn Di O. *Warm temp. regions* . . **D. sanguinalis**
+ Nerves of the lower lemma quite smooth, though sometimes hairy or beset with long glassy bristles 4

4. Racemes diverging from a central axis, this seldom exceeding the longest raceme, delicate, the spikelets loosely imbricate; spikelets 1.5–2.1mm long. Sand. Sa. *NE, E & southern Africa* . . . . **D. velutina**
+ Racemes digitate or with a short central axis in robust specimens, the spikelets closely imbricate; spikelets 2.5–3.3mm long. Damp and cultivated soils. Mp Nv Nn O R (Sa). *Tropics* . . . . . . **D. ciliaris**

90. *Pennisetum*

1. Inflorescence reduced to a cluster of 2–4 subsessile spikelets enclosed in the uppermost sheath, with long protruding filaments and stigmas. Cultivated for fodder (Kikuyu grass). Mma*. *Trop. Africa, but widely introduced* . . . . **P. clandestinum**
+ Inflorescence a spike-like panicle, conspicuously exserted 2

2. Involucres persistent, usually stipitate, the bristles plumose or glabrous; lemmas usually pubescent on the margins. Annual. Cultivated cereal (Bulrush millet). Nv* Nn* (Da*) O* Sa*. *Cult. in the tropics* . . . . **P. glaucum**

+ Involucres readily deciduous, the bristles ciliate or plumose; lemmas glabrous or almost so; wild plants 3

3. Stout annual 2m or more high; upper lemma indurated and smooth and shiny below, membranous above. Inflorescence often deeply suffused with purple. Weed of cultivation. Nv. *Trop. Africa and the foothills of the Saharan mountains* . . . . ***P. violaceum***
+ Perennials, seldom exceeding 1m in height; upper lemma of uniform texture 4

4. Bushy plant with woody stems branched throughout; leaves glaucous, tightly inrolled; sheaths usually inflated, often longer than the blade, those subtending branches characteristically shedding the blade and becomong loose and brown. Sandy desert. Nv Dl Dg Di O S R. *N Africa to India* . . . . ***P. divisum***
+ Tufted herbaceous plants, if branched and woody then only at the base and sheaths not as above 5

5. Leaves tightly inrolled, rigid, 1–2.5mm wide, the midrib noticeably thickened, appearing as a deep groove or broad pale band on the upper (inner) surface. Tough glaucous plant forming dense tussocks. Deserts, but often grown in gardens. Mma Nv (Di) S (Sa). *E trop. and northern Africa to SW Asia* . . . . ***P. setaceum***
+ Leaves flat, folded or loosely inrolled, 3–15mm wide (rarely less), the midrib not at all thickened 6

6. Inflorescence elongated, 8–30cm long, loose and often interrupted below; spikelets 4.5–6.5mm long, 1–3 in each involucre; lemmas setaceously acuminate, 5-nerved. Rocky deserts. Dg Da S. *N Africa to northern India and C Asia* . . . . ***P. orientale***
+ Inflorescence ovoid to subspherical, 5–10cm long, very dense; spikelets 9–12.5mm long, solitary in each involucre; lemmas acute, 7–9-nerved. Cultivated ornamental, sometimes escaping. (Mma*) Nv*. *NE trop. Africa and Arabia* . . . . ***P. villosum***

91. *Cenchrus*

1. Bristles of the involucre retrorsely barbellate, tenaciously clinging to clothing or fur 2
+ Bristles of the involucre antrorsely scaberulous, not clinging 3

2. Inner bristles connate only at the base to form a shallow disc. Introduced weed. Nv* (Dl*). *Trop. Africa to India* . . . . . ***C. biflorus***
+ Inner bristles (spines) fused for about half their length to form a cup. Introduced weed. Nv*. *Warmer parts of the New World* . . . ***C. echinatus***

3. Inner bristles (spines) 2–3mm long, rigid, flattened, connate for ½ to ⅔ their length to form a cup. Sand. (Da) O (R) Sa. *E trop. Africa to India* . . . . ***C. setigerus***
+ Inner bristles flexuous, filiform above, 6–16mm long 4

4. Inner bristles united only at the base to form a shallow disc 0.5–1.5mm in diameter, occasionally connate for up to 0.5mm above its rim. Wadis and areas of cultivation. Mma (Mp) Nv Nn (Dl Dg) Di S (R) Sa. *Trop. and southern Africa to India* . . . . ***C. ciliaris***
+ Inner bristles connate for 1–2.5mm above the rim of the basal disc, forming a cup. Sand. Nv (Da) Di (O) R Sa. *E trop. Africa to India* . . . . ***C. pennisetiformis***

XVII. ARUNDINELLEAE

92. *Danthoniopsis*

Densely tufted perennial, the stems branched below; leaves short, rather broad, flat and rather stiff, with conspicuous white thickened margins; spikelets in groups of three. Rocky hillsides and sandy wadis. (Da) Sa. *NE Africa and Arabia* . . . . **D. barbata**

XVIII. ANDROPOGONEAE

1. Spikelets unisexual, with male and female in separate inflorescences or in different parts of the same inflorescence 2
+ Spikelets, or at least some of them, bisexual 3

2. Racemes unisexual, the female spikelets either in a simple raceme or in rows on a thick woody cob, the male in an ample panicle . . **108. Zea**
+ Racemes bisexual, the female spikelets below, completely enclosed in a metamorphosed leaf-sheath (cupule) which takes the form of a spherical or ovoid bony bead-like structure, the male spikelets protruding from its tip . . . . **109. Coix**

3. Rhachis-internodes and pedicels slender, sometimes thickened upwards but then the upper lemma awned 4
+ Rhachis-internodes and pedicels stout, thickening upwards; upper lemma awnless 15

4. Spikelets of a pair similar, both fertile 5
+ Spikelets of a pair differing in shape and sex 8

5. Inflorescence comprising single axillary racemes . . **96. Pogonatherum**
+ Inflorescence a panicle 6

6. Raceme-rhachis fragile; one spikelet of a pair sessile . . **93. Saccharum**
+ Raceme-rhachis tough; both spikelets of a pair pedicelled 7

7. Panicle loose; glumes tough . . . . **94. Miscanthus**
+ Panicle contracted, spike-like; glumes membranous . . **95. Imperata**

8. Inflorescence a panicle with elongated central axis 9
+ Inflorescence of single or subdigitate racemes 11

9. Lower glume of sessile spikelet dorsally compressed . . **97. Sorghum**
+ Lower glume of sessile spikelet laterally compressed 10

10. Raceme composed of several to many spikelet-pairs; awns slender or inconspicuous . . . . **98. Vetiveria**
+ Raceme reduced to a triad of one sessile and two pedicelled spikelets; awns prominent . . . . **99. Chrysopogon**

11. Lower glume of sessile spikelet 2-keeled; callus more or less inserted in the hollowed internode-tip 12
+ Lower glume of sessile spikelet usually rounded without keels . . 13

12. Racemes not deflexed, borne upon unequal terete raceme-bases; leaves not aromatic . . . . **101. Andropogon**
+ Racemes usually deflexed at maturity and borne upon subequal flattened raceme-bases; leaves nearly always aromatic **102. Cymbopogon**

13. Upper lemma 2-toothed, awned . . . . **103.** *Hyparrhenia*
  + Upper lemma entire 14

14. Racemes with 2 large pairs of homogamous spikelets at the base forming an involucre . . . . **104.** *Themeda*
  + Racemes without pairs of homogamous spikelets at the base
    . . . . **100.** *Dichanthium*

15. Pedicels fused to the adjacent raceme-internode . . . **106.** *Hemarthria*
  + Pedicels free from the internodes 16

16. Callus of sessile spikelet obtuse to acute, with oblique articulation-scar; lower glume of sessile spikelet with pectinate margins . . **105.** *Elionurus*
  + Callus of sessile spikelet truncate, with transverse articulation reinforced by a central peg; lower glume of sessile spikelet without pectinate margins . . . . **107.** *Lasiurus*

93. *Saccharum*

1. Axis of panicle glabrous to pubescent; panicle up to 1m long or even more; spikelets up to 4mm long; callus-hairs off-white, shorter and scantier than in *S. spontaneum;* leaves up to 40mm wide. Cultivated grass (Sugar-cane). (Nv*) Nn*. *Cult. in the tropics* . . . . *S. officinarum*
  + Axis of panicle hirsute; panicle 25–60cm long; spikelets 3.5–7mm long; callus-hairs silky, white, 2–3 times as long as the spikelet; leaves 5–15mm wide; wild grass up to 5m high (*S. spontaneum*) 2

2. Leaves becoming petiolate towards the base, the lamina gradually reduced to a narrow wing on either side of the midrib, up to 7.5mm wide; ligule triangular. Along water-courses and on the margins of cultivated areas. Mma Nv S. *Trop. and warm temp. Asia*
    . . . . *S. spontaneum* subsp. *spontaneum*
  + Leaves with lamina extending to the base of the blade, 5–15(40)mm wide; ligule crescent-shaped. Along water-courses and on the margins of cultivated areas. Mma (Md Mp) Nv Nn (Dg Da Di O S R). *Trop. and northern Africa; Syria* . . . . *S. spontaneum* subsp. *aegyptiacum*

94. *Miscanthus*

Tussock-grass 2–3m high; inflorescence of numerous silky white racemes; leaves with horizontal bands of white variegation. Cultivated ornamental. *Native of E & SE Asia* . . . . **M. sinensis** var. **zebrinus**

95. *Imperata*

Aggressively rhizomatous perennial with dense silky panicle whose component racemes are scarcely distinguishable amongst the long silvery hairs. Watersides and irrigated land where it can become a serious weed. (Mma) Mp Nv Nn Dl Dg (Da) Di O S R. *Trop. and warm temp. Old World; S America* . . . . *I. cylindrica*

96. *Pogonatherum*

Tufted perennial with stiff wiry stems up to 60cm high. Cultivated as an ornamental (Dwarf bamboo). *Native of trop. Asia and Australia*
    . . . . *P. paniceum*

97. *Sorghum*

1. Perennial with well developed rhizomes. Once cultivated, but now an uncommon weed of rice. (Mma) Mp Nv Nn (O). *Mediterranean region to India* . . . . *S. halepense*
+ Annuals or short-lived perennial without rhizomes 2

2. Racemes tough or tardily disarticulating 3
+ Racemes fragile, readily disarticulating at maturity 4

3. Grain large, commonly exposed by the gaping glumes; sessile spikelets persistent. Cultivated cereal (Sorghum, Guinea corn). (Mma* Md*) Nv* Nn* O*. *Cult. in the tropics* . . . . *S. bicolor*
+ Grain enclosed by the glumes; sessile spikelets persistent or tardily deciduous. Subspontaneous or cultivated for fodder (Sudan grass). (Mma) Mp Nv Nn (O). *Cult. in the tropics* . . . . . . *S. x drummondii*

4. Sessile spikelet lanceolate; panicle long and narrow, scanty, 15–60 x 1–5cm. Damp sandy soils and margins of cultivation. Mma (Mp) Nv Nn Di O S. *W trop. Africa eastwards to Sudan* . . . . *S. virgatum*
+ Sessile spikelets ovate; panicle very large and full, 20–60 x 10–25cm. Unconfirmed; probably recorded in error for *S. x drummondii*. (Nv Nn). *Africa to India and Australia* . . . . *S. arundinaceum*

98. *Vetiveria*

Coarse perennial forming large clumps from stout aromatic rhizomes, up to 2m high; spikelet with shortly spiny lower glume. Introduced and cultivated for making screens. Nv*. *Native of Pakistan to SE Asia*
. . . . *V. zizanioides*

99. *Chrysopogon*

Tufted perennial with branched woody base and plumose awns. Rocky ground. Sa. *NE. Africa and Arabia* . . . . *C. plumulosus*

100. *Dichanthium*

Lower glume of sessile spikelet pitted; raceme solitary. Densely tufted perennial with basal sheaths compressed and keeled. Sandy and stony deserts. Nv Nn Dg (Da) Di (S) R Sa. *N & E Africa to India* . . *D. foveolatum*
Lower glume of sessile spikelet not pitted; racemes digitate or subdigitate. Densely tufted perennial with knotty base. Canal-banks, deep black soils and areas of cultivation. Mma Nv Nn (Dg Da) Di O S. *Trop. Africa to Indonesia* . . . . *D. annulatum*

101. *Andropogon*

Tussocky perennial with silky-pubescent basal sheaths. S. *Mediterranean region, Africa, Arabia and SE Asia* . . . . *A. distachyos*

102. *Cymbopogon*

1. Lower glume of sessile spikelet with a deep median groove from the middle downwards, corresponding to a keel on the inside; leaves rounded to cordate at the base. Cultivated for its aromatic oil (Palmerosa oil). *Cult. in the tropics* . . . . *C. martinii*

+ Lower glume of sessile spikelet flat or concave on the back; leaves narrow or attenuate at the base 2

2. Lower glume of sessile spikelet flat on the back, often wrinkled, the keels narrowly winged 3
+ Lower glume of sessile spikelet deeply concave on the back, the keels wingless 5

3. Sessile spikelet awnless. Cultivated for its aromatic oil (Lemon-grass oil). *Cult. in the tropics* . . . . *C. citratus*
+ Sessile spikelet awned 4

4. Lower glume of sessile spikelet narrowly lanceolate, usually nerveless between the keels. Cultivated for its aromatic oil (Oil of Malabar). *Cult. in India and Indo-China* . . . . *C. flexuosus*
+ Lower glume of sessile spikelet elliptic-lanceolate, usually 2- to 3-nerved between the keels. Cultivated for its aromatic oil (Citronella). *Native of Africa and India* . . . . *C. nardus*

5. Lowermost pedicel slender, not swollen. Cultivated as a condiment and for medicinal purposes. *Native of India and Pakistan* **C. jwarancusa**
+ Lowermost pedicel swollen, more or less barrel-shaped (*C. schoenanthus*) 6

6. Lower glume of sessile spikelet glabrous; inflorescence loose with racemes 20–30mm long; spatheole 23–30mm long. Deserts. Di. *N of the Sahara from Morocco to Somalia and Arabia*
. . . . *C. schoenanthus* subsp. *schoenanthus*
ı Lower glume of sessile spikelet pubescent; inflorescence dense with racemes 10–20mm long; spatheole 13–20mm long. Sandy and stony deserts. (Da) Sa. *S of the Sahara from Mauritania to Ethiopia*
. . . .*C. schoenanthus* subsp. *proximus*

103. *Hyparrhenia*

Tufted rhizomatous perennial 30–60(100)cm high; raceme-bases filiform, the upper much longer than the lower. Deserts and rocky slopes. Mma Nv Dg Di S (R) Sa. *Mediterranean region to southern Africa and SW Asia*
. . . . **H. hirta**

104. *Themeda*

Sessile spikelet rufously pubescent, awnless. Cultivated as an ornamental in El Saff in the 1950s, but probably no longer there. *Native of India and SE Asia* . . . . **T. villosa**
Sessile spikelet glabrous except for the pubescent tip, with an awn 2.5–7cm long. Dry stony ground. Nv (Di). *Trop. and subtrop. Old World*
. . . . **T. triandra**

105. *Elionurus*

Annual, deeply suffused with red either in the inflorescence or over the whole plant; lower glume of sessile spikelet with pectinate margins below, each of the blunt teeth bearing a tuft of long hair; tip of glume drawn out into two long acuminate tails. Rocky ground. (Da R) Sa. *Trop. Africa to NW India* . . . . **E. royleanus**

106. *Hemarthria*

Stoloniferous perennial with sprawling stems rooting at the nodes below; raceme-internodes fused to the adjacent pedicel. Ditches, rice-fields and other wet places. Mma Mp Nv Nn. *Italy and Turkey southwards to the Cape; isolated records in trop. Asia* . . . . **H. altissima**

107. *Lasiurus*

Tufted perennial from a thick woody rhizome; racemes white silky-villous. Sandy, stony and rocky soils. (Mma) Nv (Dl Dg) Di S R (Sa). *E trop. Africa to NW India* . . . . **L. scindicus**

108. *Zea*

Female spikelets solitary, embedded in the hardened rhachis of a single raceme, this disarticulating at maturity. Cultivated cereal (teosinte). Nv*. *Originally from Mexico* . . . . **Z. mexicana**
Female spikelets paired, gathered together in many rows on a thick woody cob. (A natural hybrid between this and teosinte has been found in Egypt). Cultivated cereal (maize, corn). Nv* (Nn*). *Cult. in the tropics and subtropics* . . . . **Z. mays**

109. *Coix*

Coarse annual up to 3m high; female spikelets completely enclosed in a bony cupule, the male exserted from its mouth. Cultivated for the cupules (Job's-tears). Nv*. *Native of trop. Asia* . . . . **C. lacryma-jobi**

Appendix 1

Taxonomic and Nomenclatural Notes

1. *Stipa lagascae* subsp. *pellita* Trin. & Rupr. is recorded from Egypt, but according to Freitag (1985) the diagnostic character - long callus 3–4mm, as compared with 2–3mm in the western Mediterranean - is part of a continuous range of variation with no strong geographical bias.

2. *Stipa parviflora* subsp. *sinaica* Chrtek & Martinovsky is at the extreme end of the range of variation of the species (Freitag 1985) with awns only 6–8(9.5)cm long, as compared with 10–13cm elsewhere. But these short-awned variants are scattered throughout the range of the species and scarcely justify its partition into subspecies.

3. *Dactylis glomerata* in the widest sense is a complex comprising a number of widespread tetraploids with enclaves, mostly in the Mediterranean region, of diploids of narrow geographical range. Individual races can be separated according to such features as stomatal dimensions, pollen size, chromosome number and average population characteristics, but since they overlap so much it is not yet clear at what rank they should be treated. Taxonomy over the whole range of the species is very uneven and no wholly satisfactory account is available. There are two distinct morphological races in Egypt. One of these is the tall, tussock forming plant typical of much of Europe and probably introduced; the other is a smaller plant with narrow panicle referred to in Egyptian literature as *D. glomerata* var. *hispanica*, but elsewhere variously treated as a subspecies or even a species. With the current state of our understanding of *Dactylis*, it is sufficient for now to acknowledge that these different populations exist without formally ascribing specfic, subspecific or varietal rank to them.

4. *Bromus* in Southwest Asia and North Africa is taxonomically in a state of flux. A number of suggestions for treatments exist but all departures from the rather old-fashioned account presented here have been ignored until things have settled down. *B. sericeus* subsp. *fallax* H. Scholz (1989), described recently on the basis of apparently aberrant material of *B. tectorum*, and a revision of the *B. fasciculatus* complex by the same author (1987) are noted, but not incorporated. A major revision of *Bromus* sect. *Genea* currently being undertaken at Edinburgh University is awaited with great interest.

5. *Elymus farctus* has several subspecies recognised in Europe and the Mediterranean region, but current treatments seem very unsatisfactory. Of all the taxa described, the least likely for Egypt, subsp. *boreali-atlanticus*, seems to be the one the local material fits best. Until these subspecies have been properly resolved and delimited, it seems better for the time being to ignore them.

Appendix 2

Checklist of Egyptian Grasses

Acrachne racemosa (*Heyne ex Roemer & Schultes*) *Ohwi*

Aegilops bicornis (*Forsskal*) *Jaub. & Spach*
  A. bicornis var. mutica Post

Aegilops crassa *Boiss.*
  A. crassa var. palaestina Eig

Aegilops kotschyi *Boiss.*
  A. kotschyi var. palaestina Eig

Aegilops longissima *Schweinf.*

Aegilops peregrina (*Hackel*) *Maire & Weiller*
  A. ovata auct. non L.
  A. variabilis Eig
  A. variabilis var. intermedia Eig & Feinbrun ex Eig

Aegilops ventricosa *Tausch*

Aeluropus lagopoides (*L.*) *Trin. ex Thwaites*
  A. brevifolius (König ex Willd.) Nees ex Steudel
  A. massauensis (Fresen.) Mattei

Aeluropus littoralis (*Gouan*) *Parl.*

Agropyron cristatum (*L.*) *Gaertner*

Agrostis stolonifera *L.*
  A. stolonifera var. scabriglumis (Boiss. & Reuter) C.E. Hubb.

Alopecurus myosuroides *Hudson*

Ammochloa palaestina *Boiss.*

Ammophila arenaria *subsp.* arundinacea *Lindb.f.*
  A. arenaria var. australis (Mabille) Hayek

Andropogon distachyos *L.*

Aristida adscensionis *L.*
  A. adscensionis var. aethiopica (Trin. & Rupr.) T. Durand & Schinz
  A. adscensionis var. pumila (Decne.) Cosson & Durieu
  A. coerulescens var. arabica Henrard

Aristida funiculata *Trin. & Rupr.*

Aristida mutabilis *Trin. & Rupr.*
  A. cassanellii Terracc.
  A. meccana Trin. & Rupr.
  A. mutabilis var. aequilonga Trin. & Rupr.

Arundo donax *L.*
  A. donax var. versicolor Stokes

Avena barbata *Pott ex Link subsp.* barbata
  A. alba Vahl

Avena barbata *subsp.* wiestii (*Steudel*) *Mansf.*

Avena fatua *L.*

Avena longiglumis *Durieu*

Avena sativa *L.*

Avena sterilis *subsp.* ludoviciana (*Durieu*) *Gill & Magne*

Avena sterilis *L. subsp.* sterilis

Boissiera squarrosa (*Banks & Sol.*) *Nevski*
   B. pumilio (Trin.) Hackel

Brachiaria deflexa (*Schum.*) *C.E. Hubb. ex F. Robyns*
   B. regularis (Nees) Stapf

Brachiaria eruciformis (*Smith*) *Griseb.*

Brachiaria leersioides (*Hochst.*) *Stapf*

Brachiaria mutica (*Forsskal*) *Stapf*

Brachiaria ramosa (*L.*) *Stapf*
   B. regularis var. *nidulans* (Mez) Täckh.

Brachiaria reptans (*L.*) *C. Gardner & C.E. Hubb.*
   Urochloa reptans (L.) Stapf

Brachypodium distachyum (*L.*) *P. Beauv.*
   Trachynia distachya (L.) Link

Briza maxima *L.*

Briza minor *L.*

Bromus aegyptiacus *Tausch*

Bromus alopecuros *Poiret*

Bromus catharticus *Vahl*
   B. unioloides Kunth

Bromus diandrus *Roth*
   B. rigens L.
   B. rigens var. *gussonii* (Parl.) T. Durand & Schinz

Bromus fasciculatus *Presl*
   B. fasciculatus var. *alexandrinus* Thell.

Bromus inermis *Leysser*

Bromus japonicus *Thunb. ex Murray*

Bromus javorkae *Pénzes*

Bromus lanceolatus *Roth*

Bromus madritensis *L.*

Bromus pectinatus *Thunb.*
   B. adoensis Steudel
   B. pulchellus Figari & De Notaris
   B. sinaicus (Hackel) Täckh.
   B. sinaicus var. *incanus* Hackel

Bromus rubens *L.*

Bromus scoparius *L.*
 B. scoparius var. *psilostachys* Hal.
 B. scoparius var. *stenantha* Stapf

Bromus sericeus *subsp.* fallax *H. Scholz*

Bromus sterilis *L.*

Bromus tectorum *L.*
 B. tectorum var. *anisanthus* Hackel

Catapodium rigidum (*L.*) *C.E. Hubb.*
 Scleropoa rigida (L.) Griseb.

Cenchrus biflorus *Roxb.*
 C. barbatus Schum.

Cenchrus ciliaris *L.*

Cenchrus echinatus *L.*

Cenchrus pennisetiformis *Hochst. & Steudel*

Cenchrus setigerus *Vahl*

Centropodia forskalii (*Vahl*) *Cope*
 Asthenatherum forskalii (Vahl) Nevski

Centropodia fragilis (*Guinet & Sauvage*) *Cope*
 Asthenatherum forskalii var. *arundinacea* (Del.) Täckh.
 A. fragile (Guinet & Sauvage) Monod

Chloris gayana *Kunth*

Chloris pycnothrix *Trin.*

Chloris virgata *Sw.*

Chrysopogon plumulosus *Hochst.*
 C. aucheri var. *quinqueplumis* (Hackel) Stapf

Coelachyrum brevifolium *Hochst. & Nees*

Coix lacryma-jobi *L.*

Cortaderia selloana (*Schultes & Schultes f.*) *Asch. & Graebner*

Corynephorus divaricatus (*Pourret*) *Breistr.*
 C. articulatus (Desf.) P. Beauv.

Crithopsis delileana (*Schultes*) *Rosch.*
 Elymus delileanus Schultes

Crypsis aculeata (*L.*) *Aiton*

Crypsis alopecuroides (*Piller & Mitterp.*) *Schrader*
 Heleochloa alopecuroides (Piller & Mitterp.) Host ex Roemer

Crypsis schoenoides (*L.*) *Lam.*
 Heleochloa schoenoides (L.) Host

Crypsis vaginiflora (*Forsskal*) *Opiz*

Cutandia dichotoma (*Forsskal*) *Trabut*

Cutandia maritima (*L.*) *Barbey*
  C. maritima var. *loliacea* (Asch.) Täckh.

Cutandia memphitica (*Sprengel*) *Benth.*

Cymbopogon citratus (*DC.*) *Stapf*

Cymbopogon flexuosus (*Nees ex Steudel*) *Watson*

Cymbopogon jwarancusa (*Jones*) *Schultes*

Cymbopogon martinii (*Roxb.*) *Watson*

Cymbopogon nardus (*L.*) *Rendle*

Cymbopogon schoenanthus *subsp.* proximus (*Hochst. ex A. Rich.*) *Maire & Weiller*
  C. proximus (Hochst. ex A. Rich.) Stapf
  C. proximus var. *sennarensis* (Hochst.) Täckh.

Cymbopogon schoenanthus (*L.*) *Sprengel subsp.* schoenanthus

Cynodon dactylon (*L.*) *Pers.*
  C. dactylon var. *villosus* Regel

Cynodon transvaalensis *Burtt Davy*

Cynosurus coloratus *Lehm. ex Steudel*

Cynosurus echinatus *L.*

Dactylis glomerata *L.*
  D. glomerata var. *hispanica* (Roth) K. Koch

Dactyloctenium aegyptium (*L.*) *Willd.*

Dactyloctenium aristatum *Link*

Dactyloctenium scindicum *Boiss.*

Danthoniopsis barbata (*Nees*) *C.E. Hubb.*

Desmazeria philistaea *subsp.* rohlfsiana (*Cosson*) *H. Scholz*
  Coelachyrum annuum Cope & Boulos
  Cutandia philistaea var. *rohlfsiana* (Cosson) Maire & Weiller

Desmostachya bipinnata (*L.*) *Stapf*

Dichanthium annulatum (*Forsskal*) *Stapf*

Dichanthium foveolatum (*Del.*) *Roberty*
  Eremopogon foveolatus (Del.) Stapf

Digitaria ciliaris (*Retz.*) *Koeler*
  D. sanguinalis var. *aegyptiaca* (Retz.) Maire & Weiller
  D. sanguinalis var. *ciliaris* (Retz.) Parl.

Digitaria nodosa *Parl.*

Digitaria sanguinalis (*L.*) *Scop.*

Digitaria velutina (*Forsskal*) *P. Beauv.*

Digitaria violascens *Link*

Dinebra retroflexa (*Vahl*) *Panzer*

Echinochloa colona (*L.*) *Link*
  E. colona var. *glauca* (Nees) Simpson
  E. colona var. *leiantha* Boiss.
  E. colona var. *repens* (Sickenb.) Simpson

Echinochloa crusgalli (*L.*) *P. Beauv.*
  E. crusgalli var. *breviseta* (Doell) Neilr.
  E. crusgalli var. *longiseta* (Doell) Neilr.
  E. crusgalli var. *mitis* (Pursh) Peterm.

Echinochloa pyramidalis (*Lam.*) *A. Hitchc. & Chase*

Echinochloa stagnina (*Retz.*) *P. Beauv.*

Ehrharta calycina *Smith*

Eleusine coracana (*L.*) *Gaertner*

Eleusine floccifolia (*Forsskal*) *Sprengel*

Eleusine indica *subsp.* africana (*Kenn.-O'Byrne*) *S. Phillips*

Eleusine indica (*L.*) *Gaertner subsp.* indica

Elionurus royleanus *Nees ex A. Rich.*

Elymus elongatus (*Host*) *Runem.*
  Elytrigia elongata (Host) Nevski
  E. elongata var. *haifensis* Meld.

Elymus farctus (*Viv.*) *Runem. ex Meld.*
  Elytrigia juncea var. *sartorii* (Boiss. & Heldr.) Täckh.

Elymus repens (*L.*) *Gould*
  Elytrigia repens (L.) *Nevski*

Enneapogon desvauxii *P. Beauv.*
  E. brachystachyus (Jaub. & Spach) Stapf

Enneapogon lophotrichus *Chiov. ex H. Scholz & P. König*

Enneapogon persicus *Boiss.*
  E. schimperianus (Hochst. ex A. Rich.) Renvoize

Enneapogon scaber *Lehm.*

Enteropogon prieurii (*Kunth*) *W.D. Clayton*
  Chloris prieurii Kunth
  C. punctulata Steudel

Eragrostis aegyptiaca (*Willd.*) *Del.*

Eragrostis aspera (*Jacq.*) *Nees*

Eragrostis barrelieri *Daveau*

Eragrostis cilianensis (*All.*) *Vign. ex Janchen*

Eragrostis ciliaris (*L.*) *R. Br.*
  E. ciliaris var. *brachystachya* Boiss.

Eragrostis japonica (*Thunb.*) *Trin.*
  E. diplachnoides Steudel

Eragrostis lepida (*A. Rich.*) *Hochst. ex Steudel*
  E. tenella auct. *non* (L.) P. Beauv. ex Roemer & Schultes

Eragrostis minor *Host*
  E. pooides P. Beauv.

Eragrostis pilosa *(L.) P. Beauv.*

Eragrostis sarmentosa *(Thunb.) Trin.*
  E. kneuckeri Hackel & Bornm.

Eragrostis tef *(Zucc.) Trotter*
  E. abyssinica (Jacq.) Link

Eragrostis tenella *(L.) P. Beauv. ex Roemer & Schultes*

Eragrostis tenuifolia *(Hochst. ex A. Rich.) Steudel*

Eragrostis tremula *Hochst. & Steudel*

Eremopoa altaica *(Trin.) Rosch.*
  Poa persica var. *diaphora* (Trin.) Asch. & Graebner

Eremopyrum bonaepartis *(Sprengel) Benth.*

Eremopyrum distans *(K. Koch) Nevski*
  E. orientale var. *lasianthum* (Boiss.) Maire

Festuca elatior *L.*
  F. arundinacea Schreber

Festuca pinifolia *(Hackel ex Boiss.) Bornm.*

Gastridium phleoides *(Nees & Meyen) C.E. Hubb.*

Halopyrum mucronatum *(L.) Stapf*

Hemarthria altissima *(Poiret) Stapf & C.E. Hubb.*
  Manisuris altissima (Poiret) A. Hitchc.

Holcus annuus *Salzm.*

Hordeum marinum *subsp.* gussoneanum *(Parl.) Thell.*
  H. marinum var. *gussoneanum* (Parl.) Täckh.

Hordeum marinum *Hudson subsp.* marinum

Hordeum murinum *subsp.* glaucum *(Steudel) Tzvelev*
  H. glaucum Steudel
  H. murinum var. *murinum* f. *fuscaspicum* Eig
  H. murinum var. *murinum* f. *violaceospicum* Eig

Hordeum murinum *subsp.* leporinum *(Link) Arcang.*
  H. leporinum Link
  H. murinum var. *leporinum* (Link) Bory & Chaub.

Hordeum spontaneum *K. Koch*

Hordeum vulgare *L.*

Hyparrhenia hirta *(L.) Stapf*

Imperata cylindrica *(L.) Raeusch.*
  I. cylindrica var. *europaea* (Pers.) Andersson.
  I. cylindrica var. *thunbergii* (Retz.) T. Durand & Schinz

Lagurus ovatus *L.*
  L. ovatus f. *nanus* Guss. ex Fiori

Lamarckia aurea (*L.*) *Moench*

Lasiurus scindicus *Henrard*
  L. hirsutus (Vahl) Boiss.

Leersia hexandra *Sw.*

Leptochloa fusca (*L.*) *Kunth*
  Diplachne fusca (L.) P. Beauv. ex Roemer & Schultes

Leptochloa panicea (*Retz.*) *Ohwi*

Leptothrium senegalense (*Kunth*) *W.D. Clayton*
  Latipes senegalensis Kunth

Lolium multiflorum *Lam.*
  L. multiflorum var. *cristatum* (Pers.) Doell
  L. multiflorum var. *gaudinii* (Parl.) Asch. & Graebner
  L. multiflorum var. *muticum* DC.
  L. multiflorum var. *ramosum* Sickenb.

Lolium perenne *L.*
  L. perenne var. *cristatum* Pers.
  L. perenne var. *ramosum* Schum.
  L. perenne var. *tenue* (L.) Hudson

Lolium rigidum *Gaudin*
  L. loliaceum (Bory & Chaub.) Hand.-Mazz.
  L. rigidum var. *compressum* (Boiss. & Heldr.) Boiss.
  L. subulatum Vis.

Lolium temulentum *L.*
  L. temulentum var. *leptochaeton* Λ. Braun
  L. temulentum var. *macrochaeton* A. Braun

Lygeum spartum *Loefl. ex L.*

Melanocenchris abyssinica (*R. Br. ex Fresen.*) *Hochst.*

Melica persica *Kunth*
  M. inaequiglumis var. *supratomentosa* Bornm.
  M. pannosa var. *schimperi* (Presl ex Steudel) Papp

Melinis minutiflora *P. Beauv.*

Melinis repens (*Willd.*) *Zizka*
  Rhynchelytrum repens (Willd.) C.E. Hubb.
  Tricholaena repens (Willd.) A. Hitchc.

Miscanthus sinensis *var.* zebrinus *Beal*

Ochthochloa compressa (*Forsskal*) *Hilu*
  Eleusine compressa (Forsskal) Asch. & Schweinf.

Oryza sativa *L.*

Oryzopsis holciformis (*M. Bieb.*) *Hackel*

Oryzopsis miliacea (*L.*) *Asch. & Schweinf.*

Panicum antidotale *Retz.*

Panicum coloratum *L.*

Panicum maximum *Jacq.*

Panicum miliaceum *L.*

Panicum repens *L.*

Panicum turgidum *Forsskal*

Parapholis filiformis (*Roth*) *C.E. Hubb.*
    Pholiurus incurvus var. *filiformis* (Roth) Täckh.

Parapholis incurva (*L.*) *C.E. Hubb.*
    Pholiurus incurvus (L.) Schinz & Thell.

Parapholis marginata *Runem.*

Paspalidium geminatum (*Forsskal*) *Stapf*

Paspalidium obtusifolium (*Del.*) *N. Simpson*

Paspalum dilatatum *Poiret*

Paspalum distichum *L.*
    P. paspalodes (Michaux) Scribner

Paspalum racemosum *Lam.*

Pennisetum clandestinum *Chiov.*

Pennisetum divisum (*J. Gmelin*) *Henrard*
    P. dichotomum (Forsskal) Del.
    P. elatum Hochst. ex Steudel

Pennisetum glaucum (*L.*) *R. Br.*
    P. niloticum Stapf & C.E. Hubb.
    P. sieberianum (Schldl.) Stapf & C.E. Hubb.
    P. spicatum (L.) Koern.
    P. typhoides (Burman) Stapf & C.E. Hubb.

Pennietum orientale *Rich.*

Pennisetum setaceum (*Forsskal*) *Chiov.*

Pennisetum villosum *R. Br. ex Fresen.*

Pennisetum violaceum (*Lam.*) *Rich.*

Phalaris aquatica *L.*

Phalaris arundinacea *var.* picta *L.*

Phalaris canariensis *L.*

Phalaris coerulescens *Desf.*

Phalaris minor *Retz.*
    P. minor var. *gracilis* (Parl.) K. Richter

Phalaris paradoxa *L.*
    P. paradoxa var. *praemorsa* (Lam.) Cosson & Durieu

Phleum pratense *L.*

Phleum subulatum (*Savi*) *Asch. & Graebner*

Phragmites australis *subsp.* altissimus (*Benth.*) *W.D. Clayton*
  P. communis var. *isiacus* Cosson

Phragmites australis (*Cav.*) *Trin. ex Steudel subsp.* australis
  P. australis var. *stenophylla* (Boiss.) Bor
  P. communis var. *stenophylla* Boiss.
  P. communis var. *striati-picta* Reichb.

Phragmites mauritianus *Kunth*
  P. communis var. *mauritianus* (Kunth) Baker

Poa annua *L.*

Poa infirma *Kunth*

Poa sinaica *Steudel*
  P. sinaica var. *vivipara* Täckh.

Pogonatherum paniceum (*Lam.*) *Hackel*

Polypogon maritimus *Willd.*
  P. monspeliensis var. *maritimus* (Willd.) Cosson & Durieu

Polypogon monspeliensis (*L.*) *Desf.*

Polypogon viridis (*Gouan*) *Breistr.*
  Agrostis semiverticillata (Forsskal) C. Chr.

Rostraria cristata (*L.*) *Tzvelev*
  Koeleria figarei De Notaris
  K. phleoides (Villars) Pers.
  K. phleoides var. *pseudolobulata* Degen & Domin
  K. phleoides var. *vivipara* Trautv.
  Lophochloa cristata (L.) N. Hylander
  L. cristata var. *pseudolobulata* (Degen & Domin) Täckh.
  L. cristata var. *pumila* (Domin) Täckh.
  L. cristata var. *vivipara* (Trautv.) Täckh.

Rostraria hispida (*Savi*) *M. Dogan*
  Koeleria hispida (Savi) DC.
  Lophochloa hispida (Savi) Täckh.

Rostraria obtusiflora (*Boiss.*) *Holub*
  Koeleria obtusiflora Boiss.
  Lophochloa obtusiflora (Boiss.) Gontch.

Rostraria pumila (*Desf.*) *Tzvelev*
  Koeleria pumila (Desf.) Domin
  K. pumila var. *glabrescens* Täckh.
  Lophochloa pumila (Desf.) Bor

Rostraria rohlfsii (*Asch.*) *Holub*
  Koeleria rohlfsii (Asch.) Murb.
  Lophochloa rohlfsii (Asch.) H. Scholz

Saccharum officinarum *L.*

Saccharum spontaneum *subsp.* aegyptiacum (*Willd.*) *Hackel*
  S. spontaneum var. *aegyptiacum* (Willd.) Hackel

Saccharum spontaneum *L. subsp.* spontaneum

Schismus arabicus *Nees*

Schismus barbatus (*L.*) *Thell.*

Schmidtia pappophoroides *Steudel*
   S. quinqueseta Benth. ex Ficalho & Hieron.

Schoenefeldia gracilis *Kunth*

Setaria italica (*L.*) *P. Beauv.*

Setaria megaphylla (*Steudel*) *T. Durand & Schinz*
   S. palmifolia auct. *non* (König) Stapf

Setaria pumila (*Poiret*) *Roemer & Schultes*
   S. glauca auct. *non* (L.) P. Beauv.
   S. lutescens (Weigel) Hubb.

Setaria verticillata (*L.*) *P. Beauv.*

Setaria x verticilliformis *Dumort.*
   S. viridis var. *ambigua* (Guss.) Cosson & Durieu

Setaria viridis (*L.*) *P. Beauv.*

Sorghum arundinaceum (*Desv.*) *Stapf*
   S. verticilliflorum (Steudel) Stapf

Sorghum bicolor (*L.*) *Moench*
   S. caudatum Stapf
   S. cernuum Host
   S. dochna (Forsskal) Snowden
   S. durra Stapf
   S. membranaceum var. *ehrenbergianum* (Koern.) Snowden

Sorghum x drummondii (*Nees ex Steudel*) *Millsp. & Chase*
   S. sudanense (Piper) Stapf

Sorghum halepense (*L.*) *Pers.*

Sorghum virgatum *Stapf*

Sphenopus divaricatus (*Gouan*) *Reichb.*
   S. divaricatus var. *ehrenbergii* (Hausskn.) T. Durand & Barratte
   S. divaricatus f. *permicranthus* (Hausskn.) Täckh.
   S. ehrenbergii Hausskn.

Sporobolus ioclados (*Nees ex Trin.*) *Nees*

Sporobolus natalensis (*Steudel*) *T. Durand & Schinz*

Sporobolus pungens (*Schreber*) *Kunth*
   S. arenarius (Gouan) Duval-Jouve

Sporobolus spicatus (*Vahl*) *Kunth*

Sporobolus wrightii *Munro ex Scribner*

Stenotaphrum secundatum (*Walter*) *Kuntze*

Stipa arabica *Trin. & Rupr.*
   S. barbata auct. *non* Desf.

Stipa capensis *Thunb.*

Stipa lagascae *Roemer & Schultes*
   S. lagascae subsp. *pellita* Trin. & Rupr.

Stipa parviflora *Desf.*
  S. parviflora subsp. *sinaica* Chrtek & Martinovsky

Stipagrostis acutiflora (*Trin. & Rupr.*) *de Winter*
  Aristida acutiflora Trin. & Rupr.
  A. zittelii Asch.

Stipagrostis ciliata (*Desf.*) *de Winter*
  Aristida ciliata Desf.

Stipagrostis drarii (*Täckh.*) *de Winter*
  Aristida drarii Täckh.

Stipagrostis hirtigluma (*Steudel ex Trin. & Rupr.*) *de Winter*
  Aristida hirtigluma Steudel ex Trin. & Rupr.

Stipagrostis lanata (*Forsskal*) *de Winter*
  Aristida lanata Forsskal

Stipagrostis multinerva *H. Scholz*

Stipagrostis obtusa (*Del.*) *Nees*
  Aristida obtusa Del.

Stipagrostis paradisea (*Edgew.*) *de Winter*

Stipagrostis plumosa (*L.*) *Munro ex T. Anderson*
  Aristida plumosa L.
  A. plumosa var. *aethiopica* Trin. & Rupr.
  A. plumosa var. *alexandrina* Trin. & Rupr.
  A. plumosa var. *brachypoda* (Tausch) Trin. & Rupr.
  A. plumosa var. *seminuda* Trin. & Rupr.

Stipagrostis raddiana (*Savi*) *de Winter*
  Aristida raddiana Savi

Stipagrostis scoparia (*Trin. & Rupr.*) *de Winter*
  Aristida scoparia Trin. & Rupr.

Stipagrostis shawii (*H. Scholz*) *H. Scholz*

Stipagrostis uniplumis (*Lichtenst.*) *de Winter*
  Aristida papposa Trin. & Rupr.

Stipagrostis vulnerans (*Trin. & Rupr.*) *de Winter*
  Aristida pungens auct. *non* Desf.
  A. vulnerans Trin. & Rupr.

Taeniatherum caput-medusae (*L.*) *Nevski*
  T. caput-medusae f. *crinitus* (Schreber) Täckh.
  T. crinitum (Schreber) Nevski

Tetrapogon cenchriformis (*A. Rich.*) *W.D. Clayton*
  T. spathaceus (Steudel) T. Durand & Schinz

Tetrapogon villosus *Desf.*

Themeda triandra *Forsskal*

Themeda villosa (*Poiret*) *A. Camus*

Tragus berteronianus *Schultes*

Tragus racemosus (*L.*) *All.*

Tricholaena teneriffae (*L.f.*) *Link*

Trichoneura mollis (*Kunth*) *E. Ekman*
  T. arenaria (Steudel) E. Ekman

Triplachne nitens (*Guss.*) *Link*

Tiraphis pumilio *R. Br.*

Trisetaria glumacea (*Boiss.*) *Maire*
  Trisetum glumaceum Boiss.

Trisetaria koelerioides (*Bornm. & Hackel*) *Meld.*
  T. koelerioides var. *aristatum* (Bornm. & Hackel) Täckh.

Trisetaria linearis *Forsskal*
  Trisetum lineare (Forsskal) Boiss.

Trisetaria macrochaeta (*Boiss.*) *Maire*
  Trisetum macrochaetum Boiss.

Triticum aestivum *L.*
  T. vulgare Villars

Triticum dicoccum (*Schrank*) *Schuebler*

Triticum durum *Desf.*

Triticum pyramidale *Percival*

Triticum turgidum *L.*

Vetiveria zizanioides (*L.*) *Nash*

Vulpia bromoides (*L.*) *Gray*

Vulpia fasciculata (*Forsskal*) *Samp.*
  V. membranacea auct. *non* (L.) Dumort.

Vulpia inops (*Del.*) *Hackel*
  V. inops var. *spiralis* Asch. & Hackel
  V. inops var. *subdisticha* (Asch. & Hackel) Barbey

Vulpia myuros (*L.*) *C. Gmelin*

Vulpia pectinella (*Del.*) *Boiss.*
  Ctenopsis pectinella (Del.) De Notaris

Zea mays *L.*

Zea mays x mexicana

Zea mexicana (*Schrader*) *Reeves & Mangelsd.*
  Euchlaena mexicana Schrader

Appendix 3

References and Further Reading

Batanouny, K.H., W. Stichler & H. Ziegler (1988). Photosynthetic pathways, distribution, and ecological characteristics of grass species in Egypt. Oecologia (Berlin) 75:539–548.

Bor, N.L. (1968) in Townsend, Guest & Al-Rawi (eds.), Flora of Iraq, volume 9 (Gramineae). Baghdad.

Bor, N.L. (1970) in Rechinger (ed.), Flora Iranica, part 70 (Gramineae). Graz, Austria.

Boulos, L. & M.N. El-Hadidi (1984). The Weed Flora of Egypt. Cairo.

Clayton, W.D. (1970) in Milne-Redhead & Polhill (eds.), Flora of Tropical East Africa, Gramineae (part 1). London.

Clayton, W.D., S.M. Phillips & S.A. Renvoize (1974) in Polhill (ed.), Flora of Tropical East Africa, Gramineae (part 2). London.

Clayton, W.D. & S.A. Renvoize (1982) in Polhill (ed.), Flora of Tropical East Africa, Gramineae (part 3). Rotterdam.

Clayton, W.D. & S.A. Renvoize (1986). Genera Graminum. HMSO.

Cope, T.A. (1985). A key to the grasses of the Arabian Peninsula. Arab Gulf Journal of Scientific Research, Special Publication No. 1.

El-Hadidi, M.N. (1980). An outline of the Planned 'Flora of Egypt.' Taeckholmia, Additional Series 1:1–12.

Freitag, H. (1985). The genus *Stipa* (Gramineae) in Southwest and South Asia. Notes from the Royal Botanic Garden, Edinburgh 42:355–489.

Scholz, H. (1987). Delimitation and classification of *Bromus fasciculatus* (Poaceae). Plant Systematics and Evolution 155:277–282.

Scholz, H. (1989). *Bromus sericeus* subsp. *fallax* (Gramineae), eine neue Unterart aus dem Vorderen Orient (Sinai). Willdenowia 19:133–136.

Sherif, A.S. & M.A. Siddiqi (1988). Poaceae, in A.A. El-Gadi (ed.), Flora of Libya. Tripoli.

Täckholm, V. (1974). Students' Flora of Egypt (ed. 2). Beirut.

Täckholm, V., G. Täckholm & M. Drar (1941). Flora of Egypt, volume 1. Cairo.

Index to Genera and Species